Selected Readings in

CHEMICAL KINETICS

Selected Readings in
CHEMICAL KINETICS

Edited, with brief explanatory notes, by

MARGARET H. BACK
Assistant Professor of Chemistry, University of Ottawa

AND

KEITH J. LAIDLER
Professor of Chemistry, and Vice-Dean,
Faculty of Science, University of Ottawa

PERGAMON PRESS
OXFORD · LONDON · EDINBURGH · NEW YORK
TORONTO · SYDNEY · PARIS · BRAUNSCHWEIG

PERGAMON PRESS LTD.
Headington Hill Hall, Oxford
4 & 5 Fitzroy Square, London W.1

PERGAMON PRESS (SCOTLAND) LTD.
2 & 3 Teviot Place, Edinburgh 1

PERGAMON PRESS INC.
44–01 21st Street, Long Island City, New York 11101

PERGAMON OF CANADA LTD.
6 Adelaide Street East, Toronto, Ontario

PERGAMON PRESS (AUST.) PTY. LTD.
20–22 Margaret Street, Sydney, New South Wales

PERGAMON PRESS S.A.R.L.
24 rue des Écoles, Paris 5ᵉ

VIEWEG & SOHN GmbH
Burgplatz 1, Braunschweig

Copyright © 1967 Pergamon Press Ltd.

First edition 1967

Library of Congress Catalog Card No. 66–30624

Printed in Great Britain by A. Wheaton & Co. Ltd., Exeter and London

CONTENTS

PREFACE   vii

Part I. Kinetic Laws

Paper 1   Excerpts from "On the Laws of Connexion between
            the Conditions of a Chemical Change and its Amount"
            by A. V. HARCOURT and W. ESSON, *Philosophical
            Transactions*, **156**, 193 (1866); **157**, 117 (1867)   3

Part II.   Molecular Kinetics

Paper 2   An excerpt, translated from the German, from S.
            ARRHENIUS's "On the Reaction Velocity of the In-
            version of Cane Sugar by Acids", *Zeitschrift für
            Physikalische Chemie*, **4**, 226 (1889)   31

Paper 3   An excerpt from W. C. McC. LEWIS's "Studies in
            Catalysis. Part IX. The Calculation in Absolute
            Measure of Velocity Constants and Equilibrium
            Constants in Gaseous Systems", *Journal of the Chemi-
            cal Society*, **113**, 471 (1918)   36

Paper 4   The first part of H. EYRING and M. POLANYI's "On
            Simple Gas Reactions", *Zeitschrift für Physikalische
            Chemie*, B, **12**, 279 (1931), translated from the German   41

Paper 5   A paper by W. F. K. WYNNE-JONES and HENRY EYRING,
            "The Absolute Rate of Reactions in Condensed
            Phases", *J. Chem. Phys.* **3**, 492 (1935)   68

Part III.   Unimolecular Reactions

Paper 6   A contribution by F. A. LINDEMANN to a Faraday
            Society Discussion on "The Radiation Theory of
            Chemical Action", published in the *Transactions of
            the Faraday Society*, **17**, 598 (1922)   93

Paper 7 The first portion of C. N. HINSHELWOOD's "On the Theory of Unimolecular Reactions", *Proceedings of the Royal Society*, A, **113,** 320 (1927)   97

Paper 8 Most of O. K. RICE and H. C. RAMSPERGER's "Theories of Unimolecular Reactions at Low Pressures", *Journal of the American Chemical Society*, **49,** 1617 (1927)   100

Part IV. Chain Reactions

Paper 9 Part I of a paper by J. A. CHRISTIANSEN, "On the Reaction between Hydrogen and Bromine", *Det. Kgl. Danske Videnskabernes Selskab. Mathematisk-fysiske Meddelelser*, **1,** 14 (1919)   119

Paper 10 A portion, translated from the German, of "The Oxidation of Phosphorus Vapour at Low Pressures", by N. SEMENOFF, *Zeitschrift für Physik*, **46,** 109 (1927)   127

Paper 11 A paper by F. O. RICE and K. F. HERZFELD, "The Thermal Decomposition of Organic Compounds from the Standpoint of Free Radicals. VI. The Mechanism of Some Chain Reactions", *Journal of the American Chemical Society*, **56,** 284 (1934)   154

Paper 12 A translation of a paper by V. A. POLTORAK and V. V. VOEVODSKY, "On a Single Chain Mechanism for the Thermal Decomposition of Hydrocarbons", published in *Doklady Academii Nauk S.S.S.R.*, **91,** 589 (1953)   171

PREFACE

IN THIS small volume we have collected twelve papers in the field of general and gas-phase kinetics. All of these papers have been important in the development of the subject, and we feel that the modern student of chemical kinetics will profit by reading the papers in their original form. The problem of selection was a somewhat difficult one. Many papers which had a profound influence at the time they were published cannot be read with much profit at the present time, except perhaps by students of the history of chemistry. We have tried to confine our selection to papers which would contribute to the student's understanding of modern aspects of the subject.

Some notes have been added to put each paper into perspective. These notes have been kept to a minimum in view of the fact that this book is a companion volume to Laidler's *Reaction Kinetics*, Vols. I and II (Pergamon Press, 1963), which provides much more detailed background information. We have added brief biographical information about the authors of the papers selected.

Three papers, those of Arrhenius, Eyring and Polanyi, and Semenoff, were translated from German. It is always a difficult task to produce a translation which faithfully reproduces the original and is also in acceptable idiomatic English; where there was a conflict we deliberately chose wording that was close to the original, so as to convey to the reader something of the style of the original article. We are grateful for help with the translation from Professor H. H. Baer and Miss O. Boshko, but ourselves accept responsibility for the final version. One paper, that of Poltorak and Voevodsky, was translated from Russian by Miss O. Boshko, and was edited by us as to style. We are most grateful to the late Professor V. V. Voevodsky for reading over the translation and suggesting changes to avoid ambiguity.

In editing these papers we have made very few changes, merely correcting obvious minor errors and misprints; we are grateful to some of the authors for help in this connection. We are glad to acknowledge that permission to reproduce has been kindly granted by:

The Royal Society (Papers 1 and 7)
Akademische Verlagsgesellschaft (Papers 2 and 4)
The Chemical Society (Paper 3)
American Institute of Physics (Paper 5)
The Faraday Society (Paper 6)
The American Chemical Society (Papers 8 and 11)
Det Kongelige Danske Videnskabernes Selskab (Paper 9)
Springer-Verlag (Paper 10)

M.H.B.
K.J.L.

PART I

Kinetic Laws

ON THE LAWS OF CONNEXION BETWEEN THE CONDITIONS OF A CHEMICAL CHANGE AND ITS AMOUNT†

A. V. HARCOURT and W. ESSON

[One of the first precise and systematic studies of the dependence of the rate of a reaction on the concentration of reactants was described by Harcourt and Esson in a series of papers the first of which was published in *Proc. Roy. Soc.* **14**, 470 (1865). The following excerpts from two later papers represent one series of experiments and give an example of the care and detail involved in their measurements and the logical methods they applied in the interpretation of the results.

At the time this paper was written Augustus George Vernon Harcourt (1834–1919) was a Student of Christ Church, Oxford, and William Esson (1838–1916) was a Fellow of New College, Oxford; the former was a chemist and the latter a mathematician. Harcourt later became Lee's Reader in Chemistry at Oxford, and Esson became Savilian Professor of Geometry at that University. In 1912, 47 years after their pioneering work on the rate laws of chemical kinetics, they again collaborated on a paper dealing with the variation of rate constants with the temperature.]

WHEN any substances are brought together under circumstances under which they act chemically one upon another, a change takes place which consists in the disappearance of a part of the original substances and the appearance of an equal weight of other substances in their place. This change continues, if the circumstances remain the same, until the whole of one of the substances taking part in it has disappeared. Its total amount is therefore ultimately determined by the amount of that substance which was originally

† Excerpts from the above article in *Philosophical Transactions*, **156**, 193 (1866); **157**, 117 (1867).

present in the smallest proportional quantity. The attainment of this limit, as will be shown, requires theoretically an infinite time, but the velocity of chemical change is so great that the practical limit of an inappreciable residue is in most cases speedily reached. Owing perhaps to this fact, chemists have been led to bestow their chief attention upon the result, and not upon the course of these changes. Occupied in investigating the relation between the reagents and the ultimate products of a reaction, and studying the chemical and physical properties of the thousand different substances thus produced, they are accustomed to regard the various conditions under which every chemical change takes place, and by which its amount is determined, chiefly as means to an end, as points to be attended to in a receipt for preparing one substance from another.

The object of the investigation which the authors have the honour of laying before the Royal Society in the following pages, has been to estimate quantitatively the relation of chemical change to some of these conditions. With this view they have selected for examination cases in which the change appeared to be of a simple character, and the conditions affecting it few in number, and capable of being defined.

Although unquestionably every chemical reaction is governed by certain general laws relating to the quantity of the substances partaking in it, their temperature and physical state, and the time during which they are in contact, yet the number of cases in which the investigation of these is practicable is extremely limited. In the first place, it must be possible both to start and terminate the reaction abruptly at a given moment. In the next, either some product or some residue of the action must be a substance for whose estimation exact and ready methods are known, that so the amount of change may be quantitatively determined. Lastly, all the conditions of the reaction must be measurable, or at least definable, and some of them susceptible of modification at will, that thus the influence of each may be examined.

After making trial of several reactions which appeared suitable, and being as often foiled by some practical difficulty in the proposed methods of investigation, we at last succeeded in devising for a very simple case of chemical change a method of investigation at once easy and exact. The reaction is that of hydric peroxide and hydric iodide,

$$H_2O_2 + 2HI = 2H_2O + I_2.$$

When solutions of potassic iodide and sodic peroxide are brought together in presence either of an acid or an alkaline bicarbonate, a gradual development of iodine takes place. If sodic hyposulphite be added to the solution it reconverts the iodine, as soon as it is formed, into iodide, but appears in no other way to affect the course of the reaction. Consequently, if the peroxide be in excess over the hyposulphite, the whole of the latter is changed by the action of nascent iodine into tetrathionate, while the amount of iodide remains always constant; and after this conversion of the hyposulphite is complete, free iodine makes its appearance in the solution.[1] The moment at which this liberation of iodine begins may be most accurately observed by the help of a little starch previously added to the liquid.

In all the experiments whose results are here recorded the same apparatus and methods were employed. The apparatus consisted of a glass cylinder about 12 inches high and 3 broad, round which, within $2\frac{1}{2}$ inches of the top, a fine line was etched: into the cylinder, through a bung closing its mouth, passed a thermometer and an inverted funnel-tube; the latter, which occupied the axis of the cylinder and reached nearly to the bottom, was connected with an apparatus for generating carbonic acid; a third hole in the bung, which served to give access to the contents of the cylinder,

[1] A solution of sodic hyposulphite may be mixed with a large volume of a dilute solution of potassic iodide and hydric sulphate or chloride without undergoing any decomposition. It is not oxidized to sulphate, nor acted upon in any way in this solution by hydric peroxide; for its decomposition is accompanied by a formation of sulphur, which even in very minute quantity would produce a perceptible opalescence in the liquid under observation. When hydric chloride has been employed to acidulate the solution, the addition of barium chloride after or during the set of experiments produces no precipitate. The quantity of sodic hyposulphite in the solution varies in each experiment from the maximum quantity to zero; the progress of the reaction is unaffected by this variation.

was ordinarily closed with a small cork. The method of perform-
ing an experiment was as follows. A quantity of water, purified
from organic matter by redistillation off potassic permanganate,
was boiled for some time to expel dissolved oxygen, and then
allowed to cool in an atmosphere of carbonic acid. When cold it
was poured into the cylinder which had previously been filled with
carbonic acid, and a current of this gas, ascending in large bubbles
from the inverted funnel, was kept passing through the liquid
until the close of the experiment. These bubbles of gas, whose
diameter is nearly half that of the cylinder, serve the purpose of
stirring the fluid constantly and to any required degree, without
causing loss or exposure to the air, and without danger to the
thermometer. Measured quantities of the standard solutions were
then introduced according to the particular experiment which was
to be made; for example, 50 c.c. of hydric sulphate and 10 c.c.
of potassic iodide, together with in all cases a few c.c. of starch.
Next, the liquid having been brought to the proper temperature,
the cylinder was placed on a level stand, and so much more water
added as would make the upper surface of the fluid exactly coincide
with the line etched upon the vessel. In every experiment the
same quantity of the remaining ingredient was taken, namely,
10 c.c. of a dilute solution of hydric peroxide;[2] thus the total
volume was in every experiment the same. Two operations,

[2] The solution employed in most of these sets of experiments was prepared
by dissolving a weighed quantity of pure sodic peroxide in water, and adding
twice the quantity of hydric sulphate required to neutralize it. The alkaline
solution of sodic peroxide, and the solution obtained by neutralizing this with
hydric sulphate, decompose slowly but perceptibly from day to day; the
addition of a second proportion of acid renders the solution almost absolutely
stable. In some sets of experiments a pure solution of hydric peroxide was
employed, which was obtained by distilling the acidified solution of sodic
peroxide. The first portions of the distillate consist of water containing but
little peroxide; as the acid liquid becomes more concentrated and the temperature rises, hydric peroxide comes over in considerable quantities, but finally
decomposition sets in, and the liquid in the retort effervesces with escaping
oxygen. About $\frac{1}{4}$ of the peroxide may thus be collected in a simple distillation.
The proportion is not much increased by distilling under a diminished atmos-
pheric pressure. It is probable that by supplying water continuously so as to
keep the fluid in the retort at that degree of concentration at which the
peroxide begins to come over in quantity, nearly the whole might be distilled.

however, had still to be performed before starting the reaction by the addition of peroxide. First, it was necessary to make sure in each case that the fluid contained no trace of any oxidizing or reducing substance. To this end the colour of the fluid was brought to the faintest possible blue by the addition, according to circumstances, of a trace of sodic hyposulphite or hydric hypochlorite. If at the end of half an hour the blue tint had neither deepened nor disappeared, it was considered that the experiment might be proceeded with. Secondly, it was necessary to add a little measure of sodic hyposulphite, the first of a series of similar measures which were to play the part already indicated in the observation of the course of the reaction. These measures needed to fulfil two somewhat incompatible conditions; they must be exactly equal, or at least must stand to one another in a known ratio; and they must be of very small volume in order that their addition might not materially augment the total volume of the fluid. In the earlier experiments a pipette of about half a cubic centimetre capacity, with a capillary tube at either end, was filled with the hyposulphite solution by means of a siphon-tube provided with india-rubber nozzle and clamp. The lower end of the pipette having been wiped and pressed gently against a pad of blotting-paper, was inserted into the cylinder through the hole in the bung so as to dip beneath the surface of the fluid. By applying the mouth to a piece of india-rubber tubing slipt over the upper end of the pipette, and alternately blowing and sucking, the charge of hyposulphite was driven and completely washed into the great body of the fluid within the cylinder. This mode of measurement proved to be susceptible of great accuracy, but it only satisfied imperfectly the second condition, that of producing no material increase in the total volume of fluid used in the experiment. It will be seen that in some of the experiments hereafter recorded as many as twenty observations of the return of the blue colour were made; the total quantity of hyposulphite that had been added at the close of such an experiment was therefore 10 c.c., causing an increase of 1 per cent in the volume of the fluid.

Subsequently a method of measurement was devised by which this source of error was avoided. It consisted in collecting single drops of a strong solution of hyposulphite under circumstances favourable to their perfect uniformity, and introducing these in succession into the cylinder. The drops were formed at the end of a siphon of which the shorter limb passed into a bottle containing the standard solution, while the longer limb, clamped firmly to a solid stand, was protected at its extremity by an outer tube which extended slightly below it and served to shield the growing drop from currents of air. The siphon was at one point so contracted as to deliver not more than five drops in a minute. The drops were received in small tubes about 6 inches long, open at both ends; in the side of each tube near one extremity a round hole had been blown such as would be made for the purpose of joining on another tube at right angles. Two forks were so placed on either side of the long limb of the siphon as that when one of these tubes had been passed through and rested upon them it lay horizontally just under the dropping-point, and could easily be so adjusted as to receive a drop through its lateral opening. When a drop had fallen, the tube containing it was withdrawn and replaced by another tube until a sufficient number of drops had been collected. An india-rubber cap was then slipped over the dropping-point to stop the flow of the liquid. The whole apparatus remained always in readiness, needing only the removal of its cap whenever it was desired to collect a fresh series of drops. The width of the bottle containing the standard solution is so great, in comparison with the quantity of solution taken for any one set of experiments, that the available length of the siphon and the rate of flow, upon whose constancy that of the drops depends, varies in no appreciable degree. Numerous determinations were made with standard iodine solution of the values of drops thus collected, and they proved to be perfectly equal. To introduce a drop into the fluid in the cylinder, the end of one of the tubes thus charged was dipped into it and moved up and down, while an active stirring was carried on by means of the bubbles of carbonic acid.

When, then, the preparations already described had been

completed and a sufficient number of tubes, each loaded with its drop, were lying in readiness, it remained to add to the contents of the cylinder a measure of hydric peroxide, and to mix it as thoroughly and as rapidly as possible with the rest of the fluid. Since, however, the addition and mixing are far from being instantaneous, an experiment was not made to date from this point, but from the moment of the first appearance of the blue colour. In order that the second at which this change occurred might be accurately noted, the cylinder was placed on a sheet of white paper in a good light, and opposite to it was stationed a clock beating seconds. The paper lay on an iron plate, one end of which was heated more or less gently by a lamp according to the temperature at which the set of experiments was to be performed and that of the surrounding air. By moving the cylinder nearer to or further from the heated end of the plate, the temperature of the fluid could be conveniently regulated. The observations were made by looking down upon the column of fluid and watching the appearance of the disk forming its upper surface, listening at the same time to the beat of the clock and counting the seconds. So suddenly does the blue shade pass over the clear and brightly illuminated disk, that a practised observer can generally feel sure as to the second in which the change begins. And where the reaction is proceeding very rapidly it would often be possible to subdivide the second. As soon as the observation had been made, a drop of hyposulphite was introduced, which speedily restores the liquid to its normal colourless condition. The time that elapses between two successive appearances of the blue colour becomes continually greater as the amount of peroxide in the solution diminishes, and finally the last measure of hyposulphite requires for its conversion more iodine than the residual peroxide can furnish, and the blue colour never returns. The values of the measure of peroxide and of the drops are readily compared by means of a standard solution of potassic permanganate. To apply this reagent to the estimation of sodic hyposulphite, it is necessary to add to the solution potassic iodide and hydric sulphate, together with a little starch. The determination may thus be performed directly with the same result as though an acidified solution of potassic iodide were first decomposed by permanganate, and the liberated iodine were then used to measure the hyposulphite. The relation between the two reactions which occur simultaneously in this determination is strictly parallel to that between the reaction of hydric peroxide and hydric iodide, which forms the subject, and the reaction of iodine and hyposulphite, which furnishes the method of our investigation. But whereas hydric peroxide acts on hydric iodide much more slowly than iodine acts on hyposulphite, hydric permanganate appears under the same circumstances to form iodine even more rapidly than it is reduced. So that in this case it is possible actually to see the double reaction, for each drop of permanganate as it enters the fluid developes for an instant the colour of iodine. But the fact of the alternate action is proved independently of this observation; for while, as has been stated, the result obtained by the addition of permanganate to the mixture of hydric iodide and hyposulphite is the same as that obtained when the two reactions are caused to occur successively, the result of the direct action of hydric permanganate on hydric hyposulphite is widely different.

The relation between the measure of peroxide and the drops of hyposulphite may also be determined in another manner. It is possible at the close of the actual set of experiments, having allowed the liquid in the cylinder to stand until the action has come practically to an end, to determine the excess of hyposulphite by means of a standard solution of iodine or permanganate, and then to determine by the same means the value of an entire drop subsequently added. Hence we know what fraction of a drop, in addition to the entire drops previously introduced, has been required to reduce the whole of the peroxide, and this quantity (the whole number and the fraction) expresses the value of the peroxide at the commencement of the experiment. If we represent by r the amount of residual hyposulphite at the close of the experiment, and by d the amount in one drop, and by n the number of drops added between the first and last appearances of the blue colour, then $\frac{r}{d}$ is the fraction of a drop which remained unacted upon, and $\dfrac{d-r}{d}$ is the fraction of a drop acted upon by the last

portion of peroxide; and putting p equal to $\dfrac{d-r}{d}$, $n+p$ is the

whole quantity acted on, and may therefore represent also the quantity of peroxide at the moment of the first observation. At the moment of the second observation the quantity of peroxide is $n-1+p$, and at the moments of subsequent observations it is successively $n-2+p$, $n-3+p$, etc., until finally at the moment of the last observation only p remains. Now the decrease of the peroxide is a measure of the amount of chemical change. Each time that the operation represented by

$$H_2O_2 + 2HI = 2H_2O + I_2$$

is performed a molecule of peroxide disappears. We may therefore regard the change by which $n+p$ parts of peroxide become $n-1+p$ parts as a definite portion of chemical change. Representing, then, the observed times by t_0, t_1, t_2, etc., t_1-t_0, t_2-t_1, etc. are the successive intervals in which as the experiment proceeds this portion of chemical change is accomplished. Now if all the conditions of the reaction could be kept constant, if it were possible to reconvert the water which is formed into hydric peroxide, as it is possible, by placing sodic hyposulphite in the solution, to reconvert the iodine which is formed into hydric iodide, then, the same event occurring always under the same conditions, the intervals t_1-t_0, t_2-t_1, etc. would be equal. But, as it is, one condition varies, namely, the quantity of hydric peroxide in the solution; and as this quantity diminishes, the amount of chemical change in a unit of time diminishes, or the time required for the accomplishment of a unit of chemical change increases. The former of these (the amount of chemical change occurring within a given time) was the quantity which we were able to determine when investigating the reaction in which hydric permanganate is gradually reduced by an excess of hydric oxalate. The latter (the time required for a given amount of chemical change) is that which we are able to measure in the experiment which we have described. Either determination provides us with the means of calculating the relation between the amount of chemical change and the varying condition, that is, the continually diminishing amount of one of the active substances.

Table I contains the results of one of our first sets of experiments. The standard solutions employed in it and in subsequent sets were (1) dilute hydric sulphate containing ·37 g in a c.c., (2) a solution of potassic iodide containing ·06 g in a c.c., (3) a solution of sodic peroxide containing ·00127 g in the same volume. Of the first of these 100 c.c. were taken and 10 of each of the others. The total volume of the solution was very nearly 1 litre. The measures of hyposulphite were such that 21·45 of them were equivalent when determined by permanganate to the measure of sodic peroxide. Before starting the experiment, by adding the solution of peroxide, half a measure of hyposulphite was introduced. At the moment, then, of the first appearance of the blue colour, from which moment the observed times in column II date, the amount of peroxide in the solution measured in drops of hyposulphite, was 20·95. The numbers in column I express the quantities of peroxide present in the solution at the observed times, those in column IV the intervals between two successive observations, and those in column III the amounts of chemical change that occurred in those intervals.

We shall find it convenient to speak of such a series of observations made after the addition of successive measures of hyposulphite as a set of experiments, and to apply the term experiment to each addition of hyposulphite and the two observations which determine the corresponding interval.

Starting, then, from the point to which our previous investigations had led us, we inquired at once whether this curve was logarithmic, that is to say, whether the amount of action had in this case varied directly with the amount of the varying active substance. The equation expressing this hypothesis has been shown to be

$$u = ae^{-\alpha x},$$

TABLE I

WEIGHTS OF SUBSTANCES TAKEN:–SODIC
PEROXIDE ·0127 g, HYDRIC SULPHATE
37·1 g, POTASSIC IODIDE ·6 g, VOLUME
OF SOLUTION 993 c.c. TEMPERATURE 17°C;
y = RESIDUE OF PEROXIDE AFTER t min;
$t'-t$ = THE TIME OF A PORTION OF
CHEMICAL CHANGE BY WHICH y IS
DIMINISHED TO y'.

I	II	III	IV
y	t	$y-y'$	$t'-t$
20·95	0·00		
19·95	4·57	1	4·57
18·95	9·37	1	4·80
17·95	14·50	1	5·13
16·95	19·87	1	5·37
15·95	25·57	1	5·70
14·95	31·68	1	6·11
13·95	38·20	1	6·52
12·95	45·23	1	7·03
11·95	52·82	1	7·59
10·95	61·12	1	8·30
9·95	70·15	1	9·03
8·95	80·08	1	9·93
7·95	91·27	1	11·19
6·95	103·88	1	12·61
5·95	118·50	1	14·62
4·95	135·85	1	17·35
3·95	157·00	1	21·15
2·95	184·53	1	27·53
1·95	223·45	1	38·92
0·95	291·18	1	67·73

where a is the amount of active substance, u the residue after a
time x, α the fraction disappearing in a unit of time, and e the
base of Napierian logarithms. To the quantity a in this equation
corresponds any of the values of y in Table I, to the quantity u
corresponds the next successive value of y in the Table, i.e. y', and
to the time x corresponds the interval $t'-t$ during which the

quantity y has diminished to the quantity y'. So that the modified form of the equation which is applicable to Table I is

$$y' = ye^{-\alpha(t'-t)}.$$

Now this may be written in the form

$$\frac{y}{y'} = e^{\alpha(t'-t)};$$

or taking the logarithms of both sides of the equation,

$$\log \frac{y}{y'} = (t'-t)\alpha \log e,$$

which expresses the fact the logarithms of the ratio of any two successive residues is proportional to the corresponding interval. For calculation it is convenient to express the equation in the deduced form

$$\log \log \frac{y}{y'} - \log (t'-t) = \log \alpha + \log \log e.$$

If, then, the differences between the corresponding values of $\log \log \frac{y}{y'}$ and $\log (t'-t)$ are found to be constant within the errors of experiment, it may be presumed that the hypothesis above stated is correct.

These values and their differences are given in Table II.

The mean of the values of $\log \log \frac{y}{y'} - \log (t'-t)$ is $\bar{3} \cdot 664$, and it will be seen that every one of the values obtained for this difference from the several experiments approximates very closely to the mean. Those which exhibit the greatest deviation on either side are $\bar{3} \cdot 668$ and $\bar{3} \cdot 660$; and it is important to ascertain whether these deviations can be accounted for by possible errors of experiment. The errors may occur (1) in the measurement of the small quantities of sodic hyposulphite, (2) in the management of the temperature of the solution, (3) in the estimation of the interval

$t'-t$, which depends upon two successive observations of the moment at which the colour of the solution changes. If, then, we suppose that the whole deviation is due to an error committed in one of these operations, the rest having been correctly performed, we find that it might result either (1) from a particular measure of hyposulphite having been one per cent smaller or

TABLE II

$\log \log \dfrac{y}{y'}$	$\log (t'-t)$	$\log \log \dfrac{y}{y'} - \log (t'-t)$
$\bar{2} \cdot 327$	$0 \cdot 660$	$\bar{3} \cdot 667$
$\bar{2} \cdot 349$	$0 \cdot 681$	$\bar{3} \cdot 668$
$\bar{2} \cdot 373$	$0 \cdot 710$	$\bar{3} \cdot 663$
$\bar{2} \cdot 395$	$0 \cdot 730$	$\bar{3} \cdot 665$
$\bar{2} \cdot 421$	$0 \cdot 756$	$\bar{3} \cdot 665$
$\bar{2} \cdot 449$	$0 \cdot 786$	$\bar{3} \cdot 663$
$\bar{2} \cdot 478$	$0 \cdot 814$	$\bar{3} \cdot 664$
$\bar{2} \cdot 509$	$0 \cdot 847$	$\bar{3} \cdot 662$
$\bar{2} \cdot 543$	$0 \cdot 880$	$\bar{3} \cdot 663$
$\bar{2} \cdot 579$	$0 \cdot 919$	$\bar{3} \cdot 660$
$\bar{2} \cdot 619$	$0 \cdot 056$	$\bar{3} \cdot 663$
$\bar{2} \cdot 663$	$0 \cdot 997$	$\bar{3} \cdot 666$
$\bar{2} \cdot 711$	$1 \cdot 049$	$\bar{3} \cdot 662$
$\bar{2} \cdot 766$	$1 \cdot 101$	$\bar{3} \cdot 665$
$\bar{2} \cdot 829$	$1 \cdot 165$	$\bar{3} \cdot 664$
$\bar{2} \cdot 902$	$1 \cdot 239$	$\bar{3} \cdot 663$
$\bar{2} \cdot 991$	$1 \cdot 325$	$\bar{3} \cdot 666$
$\bar{1} \cdot 103$	$1 \cdot 440$	$\bar{3} \cdot 663$
$\bar{1} \cdot 255$	$1 \cdot 590$	$\bar{3} \cdot 665$
$\bar{1} \cdot 495$	$1 \cdot 831$	$\bar{3} \cdot 664$

larger than the rest, or (2) from the temperature having been $0 \cdot 13°$ too high or too low, or (3) from an error of three seconds having been made in measuring an interval of five minutes. The second of these errors we may perhaps pronounce impossible: the fluctuations of the temperature of the solution seldom exceed $0 \cdot 05°$, and by balancing a small oscillation on one side of the degree line by a similar oscillation on the other, the mean thermometric error during an interval may generally be reduced to a much smaller quantity. But neither of the other errors is such as might not possibly occur in one or two out of a large number of measurements and observations. It is, however, most probable that the maximum deviations from the mean result are due, not to any single experimental error, but to the simultaneous occurrence of two or more errors in the same direction. For example, it may happen (and in eleven experiments it is an even chance that the case will occur) that the measure of hyposulphite is less than the mean, the temperature of the solution too high, the first observation made too late, and the second observation too soon. All these errors conspire to make the experiment in which they occur give

too high a number for $\log \log \dfrac{y}{y'} - \log (t'-t)$. And such a diver-

gence as that in the experiment which gives for the value of this difference $\bar{3} \cdot 668$ instead of the mean $\bar{3} \cdot 664$, would occur if the measure of hyposulphite were a fifth per cent smaller than usual, the temperature $0 \cdot 025°$ too high, and the observed interval one second too small. Now all these errors are probable experimental errors. Hence it appears that within the limits of experimental error the numerical results here obtained accord with the hypothesis before stated. In the case of this reaction, it appears that the amount of chemical change occurring at any moment is proportional to the amount of peroxide present in the solution.

In Table III the numbers obtained in several sets of experiments are similarly compared with those calculated from equations of the same form. The sets of experiments here given are selected out of a large number equally accordant with theory, with a view to illustrate the variety of circumstances under which this reaction conforms to the law which has been enunciated. For the comparison of different sets of experiments, it will be convenient to describe each solution by stating its total volume in cubic centimetres, and how many millionths of a gramme of the several ingredients it contained in a cubic centimetre. The conditions of each set of experiments are enumerated at the head of the columns which contain the intervals actually observed and those calculated from the theoretical equation. The value of α for each set of experiments is put at the head of the column which contains the calculated intervals.

The discrepancy between the observed and calculated intervals in the earlier experiments of the set made at 0°C depends upon the difficulty which was experienced in managing the temperature. If it rose at all the rate of change was of course increased, and if it fell it was increased also by the separation of some of the water from the acid solution in the form of ice. With this exception it will be seen that the calculated and observed intervals agree very closely. Hence we conclude that whether the solution contains in each c.c. 746 millionths of a gramme of hydric sulphate, or 150 times that quantity, 604 millionths of a gramme of potassic iodide or 9 times that quantity, or whether hydric chloride or hydro-sodic carbonate be substituted for hydric sulphate, whether the temperature be 0° or 50°C, and whether the portion of change require for its accomplishment intervals of one or two minutes, or intervals of half an hour or an hour, this reaction still conforms to the law that the amount of change is at each moment proportional to the amount of changing substance.

In each set of experiments we commence with a system which contains elements capable of undergoing a certain quantity of change. We may express this by saying that there exists at starting a certain amount of potential change. As time elapses this potential change gradually becomes actual. From this point of view the change occurring in the system is analogous to the motion of a heavy body falling freely, which at the commencement of its motion has a certain amount of potential energy capable of being transformed into actual energy. As the body falls the potential energy gradually becomes actual. Each experiment supplies data for the determination of the following quantities:—

(1) The initial potential change.
(2) The final potential change.
(3) The actual change.
(4) The time during which the actual change has occurred.

TABLE III

Volume, 993 c.c. Temperature, 30° Hydric sulphate, 746 Hydric iodide, 1950 Sodic peroxide, 74 Sodic hyposulphite (one measure), 13·3		Volume, 694 c.c. Temperature, 17° Hydro-sodic carbonate, 4760 Potassic iodide, 5180 Sodic peroxide, 35·7 Sodic hyposulphite (one measure), 10·7		Volume, 993 c.c. Temperature, 0° Hydric sulphate, 18700 Potassic iodide, 1208 Sodic peroxide, 59·5 Sodic hyposulphite (one measure), 10·9	
Intervals, $t' - t$		Intervals, $t' - t$		Intervals, $t' - t$	
Observed	Calculated, $a = ·0242$	Observed	Calculated, $a = ·085$	Observed	Calculated, $a = ·0043$
2·42	2·40	1·17	1·17	14·27	13·8
2·50	2·55	1·30	1·30	14·77	14·6
2·71	2·72	1·45	1·46	31·38*	32·4
2·94	2·91	1·65	1·66	17·87	18·1
3·10	3·13	1·93	1·93	19·62	19·6
3·40	3·39	2·33	2·32	21·45	21·4
3·68	3·69	2·88	2·89	23·65	23·6
4·07	4·05	3·78	3·83	26·37	26·3
4·52	4·49	5·70	5·71		
5·02	5·04	11·73	11·54		
5·77	5·74				
6·71	6·67				
7·92	7·97				
9·77	9·88				
12·91	13·00				
19·00	19·00				
37·00	36·70				

* Double interval.

TABLE III (cont.)

Volume, 993 c.c.
Temperature, 30°
Hydric sulphate, 112000
Potassic iodide, 604
Sodic peroxide, 34·9
Sodic hyposulphite (one measure), 21·9

Intervals, $t' - t$

Observed	Calculated, α = ·0949
2·31	2·32
2·98	2·98
4·15	4·17
7·01	7·00
30·38	30·27

Volume, 993 c.c.
Temperature, 30°
Hydric chloride, 13900
Potassic iodide, 604
Sodic peroxide, 27·5
Sodic hyposulphite (one measure), 14·1

Intervals, $t' - t$

Observed	Calculated α = ·0268
6·53	6·56
8·05	7·96
10·13	10·13
13·95	14·00
22·42	22·60
67·08	66·84

Volume, 993 c.c.
Temperature, 50°
Hydric sulphate, 18700
Potassic iodide, 1208
Sodic peroxide, 37·8
Sodic hyposulphite (one measure), 18·4

Intervals, $t' - t$

Observed	Calculated, α = ·131
1·00	1·01
1·18	1·17
1·37	1·38
1·70	1·70
2·20	2·19
3·08	3·09
5·33	5·32

The relation existing between these quantities has been found to be of such a nature that the ratio of the initial and final potential changes in a given system depends only upon the time of the actual change, so that if this time is constant the ratio is constant; and since the actual change is simply the difference between the initial and final potential changes, it follows that for equal intervals of time the actual change is proportional to the initial potential change. Now if we could construct a system in which the potential change remained constant, it is clear that the actual change would proceed at a uniform rate, depending upon the quality of the system and proportional to the constant potential change. In all the systems upon which our experiments have been made the potential change varies, so that we are not able directly to observe this uniform rate, but we can obtain its value indirectly in the following way.

Suppose the time of actual change to be so small that its rate may be considered uniform during that time, the actual change will be so small that the initial and final potential changes may be considered to be equal; in other words, the potential change will be constant. The ratio of the small actual change to the time of its occurrence will thus represent the uniform rate of actual change when the potential change remains constant. The equation which connects the initial and final potential changes y, y' with the time of actual change has been found to be

$$\frac{y}{y'} = e^{\alpha(t' - t)},$$

whence we obtain

$$-\frac{dy}{dt} = \alpha y.$$

Now $-dy$ is the actual change which occurs during the time dt, and from what is stated above the ratio of these small quantities is the uniform rate of actual change when the potential change y remains constant. It follows therefore that in a given system, in which there exists a constant quantity of potential change y, the uniform rate of actual change is αy. Or since α is a constant for the

given system, the rate of actual change is proportional to the
potential change. If the unit of time is one minute, α represents the
fraction of the potential change which is converted into actual
change in one minute.

> [Following this, further experiments are described in which the effect on
> the rate of a variation in iodide concentration was studied under various
> conditions. The final conclusion reached is that the amount of change
> varies directly with both the concentration of peroxide and of iodide.
> Finally a theoretical discussion of the results is outlined in an Appendix
> in which the rate expressions for first- and second-order reactions and
> various cases of more complex reactions are clearly set forth.]

APPENDIX

*containing a Theoretical Discussion of some cases of Chemical
Change.*

By WILLIAM ESSON, *M.A., Fellow of Merton College, Oxford*

The most simple case of chemical change occurs in a system in
which a single substance is undergoing change in presence of a
constant quantity of other substances, and at a constant temperature. A practical constancy of the other substances is obtained by
having them present in large excess; for any change produced in
their amount by reason of the change of the single substance is
infinitesimal in comparison with their original amount, and its
effect on the system may therefore be neglected.

By a "system" is meant a unit of volume in which given quantities of substances are present; these quantities are called "elements of the system;" "a system in which a single substance is
undergoing change," is a system in which the variation of the
other substances does not affect the change of the single substance.

It has been ascertained by experiment that the residue y of the
substance undergoing change in a system of this kind, is connected with the time x during which the change has been proceeding, by the following equation,

$$y = ae^{-\alpha x}, \tag{1}$$

a being the quantity of the substance in the system at the commencement of the change, and a a constant, the meaning of which may be thus determined; differentiating (1) and eliminating x, we have

$$\frac{dy}{dx} = -ay. \tag{2}$$

Now $-\dfrac{dy}{dx}$ is the amount of substance which disappears in a unit

of time at the time x, when y is the quantity of substance present in the system, and the equation (2) expresses the law that "the amount of change in a unit of time is directly proportional to the quantity of substance;" following the analogy of the motion of a

material particle, we may call $\dfrac{dy}{dx}$ the rate or velocity of chemical

change, and the law may be thus stated:— "The velocity of chemical change is directly proportional to the quantity of substance undergoing change."

The constant a expresses the fraction of the substance which is changed in a unit of time; this fraction depends upon the other elements of the system, and upon its physical conditions, such as temperature, density, etc. By varying each of these conditions in succession, it is possible to determine a as a function of them, and to predict the progress of the chemical change of a single substance, from its commencement to its completion, under any assignable conditions.

Let us first take the case in which the chemical change consists of the reaction of two substances, neither of which is present in the system in great excess. In the discussion of this case we shall assume the general truth of the law of variation of the rate of chemical action, which has been derived from experiments in which the constancy of all the elements but one has been secured by taking them in excess. In fact we shall assume that the truth of the law depends only upon the constancy of the elements, and not upon their excess. Since, then, the velocity of change of each substance is proportional to its quantity when the quantity of the other is constant, it follows that the velocity of change is proportional to the product of the quantities when both vary. Let a, b be the number of equivalents of the substances present in the system at the commencement of the reaction, z the number of equivalents of each which has disappeared during a time x, then $a-z$, $b-z$ are the number of equivalents remaining at the end of that time; hence

$$\frac{dz}{dx} = n(a-z)(b-z), \tag{3}$$

the solution of which is

$$\log \left(1 - \frac{z}{a}\right) - \log \left(1 - \frac{z}{b}\right) = n(a-b)x, \tag{4}$$

an equation for determining the amount of chemical change, in this case, after the lapse of a given time.

When the substances are originally present in equivalent quantities, $a = b$, and (3) becomes

$$\frac{dz}{dx} = n(a-z)^2, \tag{5}$$

the solution of which is

$$z = a \frac{nax}{nax + 1}. \tag{6}$$

The equation connecting the residue y with the time is in this case

$$y = \frac{a}{nax + 1}; \tag{7}$$

and if at the commencement of the reaction the substances had been present in infinitely large quantities,

$$y = \frac{1}{nx}. \tag{8}$$

The curve (6), which expresses the reaction of two substances originally present in equivalent quantities, is a rectangular hyperbola, and when the original quantities are infinite, the residue varies inversely as the time.

Let us suppose that at the commencement of the reaction there are present a equivalents of a substance A, which during the course of the reaction is gradually changed into an equivalent quantity of a substance B, and that B reacts with a substance C of which a equivalents are originally present; also let u be the number of equivalents of A which remain after an interval x, and v the number of equivalents of B which remain after the same interval; then, since the velocity of diminution of u is proportional to its quantity, and the velocity of diminution of v proportional to the product of its quantity into the quantity of c, and the velocity of increase of v equal to the velocity of diminution of u, we have the following equations,

$$\frac{du}{dx} = -\beta u, \tag{9}$$

$$\frac{dv}{dx} = -av(u+v)+\beta u. \tag{10}$$

The solution of (9) is

$$u = ae^{-\beta x}; \tag{11}$$

so that if the residue of u could be measured separately from that of v, the rate of change of u into v could be determined, but in the actual experiments u and v are determined together, and the relation between the total residue $y(=u+v)$ and the duration of the reaction x is consequently very complex.

Adding (9) and (10), we have,

$$\frac{dy}{dx} + avy = 0; \tag{12}$$

substituting for dx from (9), and for v its value $y-u$, we obtain the equation

$$\frac{dy}{y^2 du} + \frac{a}{\beta y} - \frac{a}{\beta u} = 0, \tag{13}$$

the solution of which is

$$\frac{a}{\beta} e^{\frac{a}{\beta}u} \left\{ c - \log u + \frac{a}{\beta}u - \frac{1}{1.2^2}\left(\frac{a}{\beta}u\right)^2 + \ldots \right\} y = 1. \tag{14}$$

If we replace for u its value $ae^{-\beta x}$, we obtain an equation connecting the residue y with the time x.

The next case to be considered is that of a system in which there are two substances undergoing change in presence of a large excess of the other elements of the system. If both substances are present in the system from the commencement of the change and are independent of each other, the velocity of diminution of each is proportional to its quantity, and their residues accord with the simple law $y = ae^{-\alpha x}$; and if both these residues are measured together, the equation of the reaction is

$$y = a_1 e^{-\alpha_1 x} + a_2 e^{-\alpha_2 x}, \tag{17}$$

a_1, a_2 being the quantities of the substances originally introduced into the system, and α_1, α_2 the fractions of them which disappear in a unit of time.

If, however, the substances are not independent, but are such that one of them is gradually formed from the other, we have a different system of equations to represent the reaction.

Let u, v be the residues of the substances after an interval x, y ($=u+v$) being the total residue actually measured at that time. Let the initial values of u and v be $u=a$, $v=0$; let αu be the rate of diminution of u due to its reaction with one of the other elements of the system, and βu its rate of diminution due to its reaction with another of the elements of the system, by means of which v is formed, and let γv be the rate of diminution of v, then

$$\frac{du}{dx} = -(\alpha+\beta)u, \tag{18}$$

whence
$$\frac{dv}{dx} = \beta u - \gamma v, \tag{19}$$

$$u = ae^{-(\alpha+\beta)x}, \tag{20}$$

$$v = \frac{a\beta}{\alpha+\beta-\gamma} \left\{ e^{-\gamma x} - e^{-(\alpha+\beta)x} \right\}, \tag{21}$$

$$y = \frac{a}{\alpha+\beta-\gamma} \left\{ \beta e^{-\gamma x} + (\alpha-\gamma)e^{-(\alpha+\beta)x} \right\}. \tag{22}$$

B

There are several particular cases of these equations which require to be considered separately.

(1) $\beta=0$. Fraction of v formed$=0$.

In this case the system of equations reduce to

$$u = ae^{-\alpha x},$$
$$v = 0,$$
$$y = ae^{-\alpha x}.$$

(2) $\gamma > \alpha$. Fraction of v decomposed in a unit of time, greater than the fraction of u decomposed in a unit of time.

In this case the last equation of the system is of the form

$$y = a_1 e^{-\alpha_1 x} - a_2 e^{-\alpha_2 x}.$$

(3) $\gamma=\alpha$. The fraction of v decomposed in a unit of time equal to the fraction of u decomposed in a unit of time.

In this case the last equation of the system reduces to the form

$$y = ae^{-\alpha x}.$$

(4) $\gamma < \alpha$. The fraction of v decomposed in a unit of time greater than the fraction of u decomposed in a unit of time.

In this case the last equation of the system is of the form

$$y = a_1 e^{-\alpha_1 x} + a_2 e^{-\alpha_2 x}.$$

It is thus possible to have all these four cases in succession in a set of experiments in which only one condition is progressively varied, provided that the variation of γ and α is such that γ is at first greater than α, but increases in a less ratio than α. Several attempts have been made to calculate equations of the form $y = a_1 e^{-\alpha_1 x} \pm a_2 e^{-\alpha_2 x}$ which should give the experimental numbers within the errors of experiment, and at the same time yield values of the fractions α, β, γ from which the law of their variation with a variable quantity of sulphuric acid could be discovered. The number and exactness of the experimental results are, however, not sufficient to enable us to extract from the complicated equation

$$y = \frac{a}{a+\beta-\gamma} \left\{ \beta e^{-\gamma x} + (a-\gamma)e^{-(a+\beta x)} \right\}$$

trustworthy values of α, β, γ, and this inexactness precludes the possibility of investigating the law of their variation when the conditions of the experiment are varied. What we can state with certainty is, that the numbers are all satisfied by equations of the forms

$$y = a_1 e^{-\alpha_1 x} - a_2 e^{-\alpha_2 x},$$
$$y = a e^{-\alpha x},$$
$$y = a_1 e^{-\alpha_1 x} + a_2 e^{-\alpha_2 x},$$

and that successive sets of numbers, obtained by varying one condition progressively, are satisfied by these successive forms of equations. These forms, and the order of their succession, are accounted for by a hypothesis for which there is considerable experimental evidence, and it is thus highly probable that the results arrived at in the above discussion give a true account of the progress of the reaction.

The law of variation of α, β, γ with the conditions of the system will probably be detected when the case in which β, γ both vanish for all conditions of the system, has been fully discussed.

A complete investigation of this case is reserved for a future communication.

PART II

Molecular Kinetics

ON THE REACTION VELOCITY OF THE INVERSION OF CANE SUGAR BY ACIDS†

S. ARRHENIUS

[The so-called "Arrhenius law", which relates the rate constant k of a reaction to the absolute temperature T, is one of the most fundamental in chemical kinetics. The relationship may be expressed as

$$k = A\,e^{-E/RT} \qquad (1)$$

where A and E are constants and R is the gas constant. Its general form appears to have been first discovered empirically by J. J. Hood (*Phil. Mag.* **6**, 371 (1878); **20**, 323 (1885)). Some significance was given to the law by a thermodynamical argument due to J. H. van't Hoff, and this is summarized by Arrhenius in the passage quoted below. In the latter part of this passage Arrhenius gives further significance to the law by considering the matter from a statistical point of view.

Svante August Arrhenius (1859–1927) was a scientist of extraordinary versatility. Born in Sweden, he studied at the Universities of Uppsala and Stockholm, and in 1884 proposed in his doctoral dissertation that electrolytes are dissociated in water; this theory was slow to gain acceptance, but won him the 1903 Nobel Prize in Chemistry. He did not work extensively on the theory of chemical kinetics; his discussion of the "Arrhenius law", which is brief and to the point, represents his main contribution in that field. He was a pioneer in the application of physico-chemical methods to the study of biological systems, and wrote a book on immunochemistry. He also published on cosmology, the causes of the ice ages, and the origin of life.

The excerpt which follows is taken from pages 230 to 234 of his paper.]

IN HIS notable book *Studies in Chemical Dynamics* van't Hoff gives a theoretically-based formulation of the influence of temperature on the rate of reaction. Thus, in the case of an equilibrium

† An extract, translated from the German, from an article in *Zeitschrift für Physikalische Chemie*, **4**, 226 (1889).

between four substances A, B, E and D (of which A and B can be formed from E and D, and vice versa) we have

$$k_1 C_A C_B = k_2 C_E C_D \tag{2}$$

in which equation C_A, etc., signify the concentrations of the substances A, etc. This relationship also says that the amounts of E and D produced in unit time from A and B are equal to the amounts of A and B formed in the same time from E and D. That is to say, the first quantity is represented by the term $k_1 C_A C_B$ where k_1 is the rate constant (the specific reaction velocity): in the same way $k_2 C_E C_D$ represents the quantity of A and B newly formed in unit time. For the system at equilibrium van't Hoff gives

$$\frac{d \log_{nat} k_1}{dT} - \frac{d \log_{nat} k_2}{dT} = \frac{q}{2T^2} \tag{3}$$

an equation that can be derived from the mechanical theory of heat and van't Hoff's law. In this equation T is the absolute temperature and q the amount of heat (in calories) which is set free when 1 gram-molecule each of A and B are converted into E and D.

The two velocity constants for two reciprocal reactions are thus related by this equation (3). "This equation", says van't Hoff, "does not provide the desired relationship between the value of k (the reaction velocity) and the temperature; it shows, however, that this relationship has the following form:

$$\frac{d \log_{nat} k}{dT} = \frac{A}{T^2} + B." \tag{4}$$

In equation (4) A and B are treated as constants. It is, however, easily seen that B can be any function, $F(T)$, of the temperature, and equation (3) can only be obtained from (4) if the $F(T)$ belonging to the two reciprocal reactions are the same. I am of the opinion, with regard to this, that there is no definite solution of this problem, since $F(T)$ can be anything at all.

It is not possible to proceed further without introducing a new hypothesis, which is in a certain sense a paraphrase of the observed facts. In order to reach such a hypothesis, which will be used throughout, the following considerations must be taken into account. The influence of temperature on the specific reaction rate is very large in that, at ordinary temperatures, the rate increases by 10 to 15 per cent for each one-degree rise in temperature. It cannot be assumed, therefore, that the increasing reaction velocity comes from the increasing frequency of collisions of the reacting molecules. According to the kinetic theory of gases, the velocity of the gas molecules changes only by about 1/6 per cent of its value for each one-degree rise in temperature and the frequency of collisions increases in the same ratio. It is difficult to say with certainty how large is the corresponding change in the case of liquids, but it is suggested that it is similar to that in gases. Even if the assumption that the velocity of the solute molecules changes by 1/6 per cent per degree is not accepted, it must at least be agreed that the difference between this value and the observed 10 to 15 per cent per degree is much too large for it to be assumed that the increase in the number of collisions of the reacting molecules is the reason for the increase in the reaction velocity with the temperature. In the same way it can not very well be assumed that the decrease of viscosity with increasing temperature is the reason for the phenomenon under consideration, for the viscosity decreases[1] by only some 2 per cent per degree. Apart from this, the increase in reaction velocity differs in a very important way from the increase with temperature in most physical quantities. Thus the change of these quantities is approximately equal for an increase in temperature of 1°, even for very different values of the temperature (e.g. 0° and 50°). On the other hand the increase (in absolute value) per degree of the reaction velocity is perhaps twice as large at 6° as at 0°, at 12° four times as large as at 0°, at 30° perhaps thirty (2^5) times as large as at 0°, etc. This fact indicates that the increase in reaction velocity with the temperature cannot be explained in terms of the change in physical properties of the reacting substances. There remains one way out. A similar extraordinarily large change in specific reaction velocity

[1] [The original paper says "increases" (*nimmt zu*) but this must be a mistake.]

(k) is to be found in another sphere. This is the change in rates of reactions which are brought about by weak bases or acids, by the addition of insignificant amounts of neutral salts (e.g. by addition of NH_4Cl to NH_3 acting upon ethyl acetate). In this case the facts are explained thus: although the amount of the actual base (NH_3) or acid does not undergo a change on the addition of neutral salt (NH_4Cl), by its addition there is a very rapid removal of free OH^- ions, which actually cause the reaction. Can it not be assumed that the reaction velocity (for example the inversion of cane sugar) also originates in this way, that the amounts of the actual reacting substances rapidly increase with temperature? One actual reacting substance is the H^+ ion of the acid employed. The concentration of this ion (for a constant amount of acid) changes only very little with temperature for strong acids like HCl (HNO_3 or HBr), increasing (by something like $0 \cdot 05$ per cent per degree at $25°$) when the temperature rises. The explanation therefore can not lie here. It must therefore be assumed, to be consistent, that the other actual reacting substance is not cane sugar, since the amount of sugar does not change with temperature, but is another hypothetical substance which is regenerated from cane sugar as soon as it is removed through the inversion.

This hypothetical substance, which we call "active cane sugar", must rapidly increase in quantity with increasing temperature (by about 12 per cent per degree) at the expense of the ordinary "inactive" cane sugar. It must be formed from cane sugar at the expense of heat (of q calories). It is further known that cane sugar in solution at high temperatures has no important properties that it does not possess at lower temperatures, as would probably be the case if the absolute amount of the hypothetical active cane sugar were increased considerably; it may therefore be assumed that the absolute amount (although it increases by some 12 per cent per degree) is still, at the highest temperatures employed, exceedingly small in comparison with the amount of the "inactive cane sugar". The amount of the latter is therefore independent of temperature. Furthermore at constant temperature the reaction velocity is approximately proportional to the amount of cane sugar. Under these conditions (constant temperature), therefore, the amount of "active cane sugar", M_a, is approximately proportional to the amount of inactive cane sugar, M_i. The equilibrium condition is thus:

$$M_a = k\, M_i \tag{5}$$

The form of this equation shows us that a molecule of "active cane sugar" is formed from a molecule of inactive cane sugar either by a displacement of the atoms or by addition of water. In the equilibrium equation there is always on each side the product of the amounts (concentrations) of the reacting substances. Since, however, the amount of water is constant (the amount of cane sugar varying) the water can also be active in the reaction without its concentration appearing in the equation. Thus for the constant k (or what is the same thing M_a/M_i) we have the equation

$$\frac{d \log_{\text{nat}} k}{dT} = \frac{q}{2T^2} \tag{6}$$

which integrates to

$$k_{T_1} = k_{T_0}\, e^{q(T_1 - T_0)/2T_0 T_1}.$$

If it is further assumed that the reaction velocity is proportional only to the amounts of the reacting substances, a similar equation can be written down for the reaction velocity corresponding to a constant amount of reacting acid:

$$\rho_{T_1} = \rho_{T_0}\, e^{\,q(T_1 - T_0)/2T_0 T_1}.$$

We have given this equation above as equation (1) and have found it to be in agreement with experiment.

STUDIES IN CATALYSIS. PART IX. THE CALCULATION IN ABSOLUTE MEASURE OF VELOCITY CONSTANTS AND EQUILIBRIUM CONSTANTS IN GASEOUS SYSTEMS†

W. C. McC. Lewis

[Over twenty-five years elapsed between Arrhenius's interpretation of the energy of activation E and the quantitative treatment of the constant A which appears in the equation $k = A e^{-E/RT}$. In 1916 M. Trautz published a paper (*Zeitschrift für anorganische Chemie*, **96**, 1 (1916)) in which he related the quantity A—now generally known as the frequency factor— to the frequency of collisions between the reacting molecules, and made calculations of the magnitudes of frequency factors. Shortly afterwards W. C. McC. Lewis, who had not seen Trautz's paper because of the First World War, published a paper in which a very similar point of view was taken. Lewis's paper is a little more explicit, and is more readable today; for that reason we include an excerpt from his paper.

At the time he published this paper William Cudmore McCullagh Lewis (1885–1956) was Brunner Professor of Physical Chemistry at the University of Liverpool. He worked on a variety of problems in physical chemistry, especially chemical kinetics, and was particularly interested in the application of physical chemistry to biological systems. His book *A System of Physical Chemistry* was at one time the standard textbook on this subject.]

IN THE case of a bimolecular reaction, such as the decomposition of hydrogen iodide in the gaseous state, the fractional number of

† An excerpt from this article which appeared in *Journal of the Chemical Society*, **113**, 471 (1918).

hydrogen iodide molecules which exist in the active state may be calculated by the aid of the expression

$$N_a/N = e^{-E/RT} \tag{1}$$

where N_a denotes the number of active molecules, N the number of passive molecules or the total number of molecules (since N_a is very small compared with N); E is the critical increment reckoned per gram-molecule, that is, the amount of energy which one gram-molecule of the substance must absorb in order to make it reactive, and R and T have their usual significance. This expression is the familiar one obtained on the basis of statistical mechanics for the distribution of molecules in a field of force. The justification for its application to the present case is furnished by the following calculation of the velocity constant of decomposition of gaseous hydrogen iodide.

In Bodenstein's experiments (*Zeitsch. physikal. Chem.*, **29**, 295, 1899), one gram-molecule was present in 22·4 litres. At 556° abs. the observed velocity constant was $9·42 \times 10^{-7}$, the unit of time being the minute. On expressing the time in seconds and the concentration in gram-molecules per litre, the velocity constant becomes $3·517 \times 10^{-7}$. This number represents the fraction decomposed per second at unit concentration.

We have now to calculate the velocity constant on the basis of the concept of active molecules, the equilibrium concentration of which is assumed to be given at all stages of the observed reaction by equation (1). From the temperature-coefficient[1] of the reaction, it is calculated that the critical increment per gram-molecule is 22,000 cals. Hence, employing equation (1), it is found that the fraction of one gram-molecule which exists in the active state at 556° abs. is $2·218 \times 10^{-9}$. If there is one gram-molecule of hydrogen iodide present in 1 litre, then this number represents the fractional number of active molecules. Since there are $6·1 \times 10^{23}$ molecules in one gram-molecule, the actual number of active molecules per litre is $1·35 \times 10^{15}$, or $1·35 \times 10^{12}$ per c.c.

[1] This is the most direct means of obtaining the critical increment. It may also be obtained from the position of the effective absorption band in the spectrum of the substance provided the data are available.

On the kinetic theory, the number of collisions per c.c. per second between like (active) molecules is given by the expression

$$\sqrt{2} . \pi . \sigma^2 . u . N_a^2$$

where N_a is the number of active molecules per c.c., u the mean velocity of translation per molecule, and σ the distance within which two molecules approach one another during a collision. Physical theory has not yet succeeded in defining σ with precision, beyond the fact that it is of the order of magnitude of the radius or diameter of the molecule. Such being the case, we shall take a mean value, 2×10^{-8} cm, and employ this in all cases examined. Naturally, this will introduce a certain error into the results, but its magnitude will not affect the general question of the verification of the method of treatment adopted.

In the case of hydrogen iodide at 556° abs., $u = 3\cdot3 \times 10^4$ cm per second. The value of N_a we have already calculated to be $1\cdot35 \times 10^{12}$ per c.c. Hence the number of collisions per c.c. per second between the active molecules is $1\cdot065 \times 10^{14}$, or the number of collisions per litre is $1\cdot065 \times 10^{17}$. At each collision between active molecules, two such molecules react. Hence the number of molecules of hydrogen iodide which react per second per litre is $2\cdot13 \times 10^{17}$. Expressing this as a fraction of one gram-molecule, we obtain $2\cdot13 \times 10^{17}/6\cdot1 \times 10^{23} = 3\cdot5 \times 10^{-7}$. This should be the velocity constant of the reaction expressed in gram-molecules per litre per second. The observed value is $3\cdot517 \times 10^{-7}$. The agreement is very satisfactory, especially in view of the possible error in σ. This calculation serves to substantiate the concept of active molecules defined in the above sense.

The foregoing calculation may be carried out in a somewhat different manner which leads directly to the differential equation expressing the reaction velocity. Thus the number of molecules which react per c.c. per second is given by:

$$2\sqrt{2}\pi \sigma^2 u N_a^2.$$

Hence the number of molecules which react per litre per second is given by:

$$2000\sqrt{2}\pi \sigma^2 u N_a^2.$$

DECOMPOSITION OF HYDROGEN IODIDE

T	$u \times 10^{-4}$	Fraction of one gram-molecule in the active state $= e^{-E/RT}$	$e^{-2E/RT}$	k calculated [equation (2)]	k observed
556°	3·30	$2 \cdot 218 \times 10^{-9}$	$4 \cdot 898 \times 10^{-18}$	$3 \cdot 5 \ \times 10^{-7}$	$3 \cdot 517 \times 10^{-7}$
575	3·356	$2 \cdot 704 \times 10^{-9}$	$1 \cdot 820 \times 10^{-17}$	$1 \cdot 319 \times 10^{-6}$	$1 \cdot 217 \times 10^{-6}$
629	3·510	$2 \cdot 244 \times 10^{-8}$	$5 \cdot 012 \times 10^{-16}$	$3 \cdot 800 \times 10^{-5}$	$3 \cdot 02 \times 10^{-5}$
647	3·559	$3 \cdot 656 \times 10^{-8}$	$1 \cdot 318 \times 10^{-15}$	$10 \cdot 23 \times 10^{-5}$	$8 \cdot 587 \times 10^{-5}$
666	3·612	$5 \cdot 970 \times 10^{-8}$	$3 \cdot 548 \times 10^{-15}$	$2 \cdot 768 \times 10^{-4}$	$2 \cdot 195 \times 10^{-4}$
683	3·657	$8 \cdot 995 \times 10^{-8}$	$8 \cdot 128 \times 10^{-15}$	$6 \cdot 421 \times 10^{-4}$	$5 \cdot 115 \times 10^{-4}$
700	3·702	$1 \cdot 337 \times 10^{-7}$	$1 \cdot 778 \times 10^{-14}$	$1 \cdot 422 \times 10^{-3}$	$1 \cdot 157 \times 10^{-3}$
716	3·744	$1 \cdot 905 \times 10^{-7}$	$3 \cdot 548 \times 10^{-14}$	$2 \cdot 87 \ \times 10^{-3}$	$2 \cdot 501 \times 10^{-3}$
781	3·912	$6 \cdot 918 \times 10^{-7}$	$4 \cdot 786 \times 10^{-13}$	$4 \cdot 04 \ \times 10^{-2}$	$3 \cdot 954 \times 10^{-4}$
1000	4·42	$1 \cdot 545 \times 10^{-5}$	$2 \cdot 371 \times 10^{-10}$	$22 \cdot 63$	—

N_a is the number of active molecules per c.c. Hence the number of active molecules per litre is $1000\ N_a$. If N_0 be the number of molecules in one gram-molecule, the number of active gram-molecules per litre is $1000\ N_a/N_0$. If we denote this by C_a, then $N_a = N_0 C_a/1000$, or $N_a{}^2 = N_0{}^2 C_a{}^2/10^6$. Hence the number of molecules which react per litre per second is:

$$2000\sqrt{2}\pi\sigma^2 u N_0{}^2 C_a{}^2/10^6.$$

The number of gram-molecules which react per litre per second is $1/N_0$ of the above quantity. That is, the number of gram-molecules which react per litre per second is

$$2000\sqrt{2}\pi\sigma^2 u N_0 C_a{}^2/10^6.$$

From equation (1), it follows that $C_a = Ce^{-E/RT}$, where C is the total concentration or number of gram-molecules of hydrogen iodide per litre. Hence the rate of the observed reaction is given by:

$$-dC/dt = 5\cdot40 \times 10^{21} . \sigma^2 . u . C^2 . e^{-2E/RT}.$$

But the rate, $-dC/dt = k_{obs} . C^2$,

where k_{obs} is the velocity constant experimentally determined. Hence, k_{obs} (in gram-molecules per litre, per second)

$$= 5\cdot40 \times 10^{21} . \sigma^2 . u . e^{-2E/RT} \tag{2}$$

By making use of the concept of active molecules, we can calculate the velocity constants of decomposition of hydrogen iodide over the temperature range corresponding with Bodenstein's observations. The table on page 39 contains the calculated and observed velocity constants expressed in gram-molecules per litre per second.

ON SIMPLE GAS REACTIONS†

H. EYRING and M. POLANYI

[Following the publication of Arrhenius's paper (Paper 2) on the significance of the temperature dependence of the rates of chemical reactions, a number of workers interested themselves in the problem of the significance of the activation energy. With the advent, in the nineteen-twenties, of the new quantum mechanics it was natural to see to what extent activation energies could be calculated by the same methods that were beginning to be used for the calculation of the energies of stable molecules.

The most successful of the early attempts of this kind was that described in the paper reproduced in part below. In it Eyring and Polanyi show that a reasonable estimate of the activation energy of a reaction involving three atoms can be obtained by applying quantum-mechanical procedures with some assistance from empirical methods. A particularly important feature of the Eyring–Polanyi work is their use of potential-energy surfaces, which provide a very valuable pictorial representation of the course of a chemical reaction.

Henry Eyring was born in 1901, and at the time this paper was written was a (U.S.) National Research Fellow in Berlin. Later he became Professor of Chemistry at Princeton University and is now Dean of the Graduate School at the University of Utah. He has made many distinguished contributions, particularly in the application of theoretical methods, to the understanding of the rates of chemical and physical processes. Perhaps his most pioneering work is represented by the present paper and by Paper 5, which deals with the application of statistical methods to the rates of chemical reactions.

Michael Polanyi was born in 1891 in Budapest, and at the time this paper was written was a member of the Kaiser Wilhelm Institut für Physikalische Chemie in Berlin. He was later Professor of Physical Chemistry at Manchester University, and in 1948 became Professor of Social Studies at that University. He has made important contributions in kinetics and in related branches of physical chemistry, and is the author of a number of books on sociology.]

---

† The first part of an article, translated from the German, which appeared in *Zeitschrift für Physikalische Chemie*, B, **12**, 279 (1931).

Abstract

A further development of London's theory of adiabatic chemical processes, especially atom reactions. The total binding energy of single pairs of atoms is determined from optical data as a function of the interatomic distance and is corrected for the coulombic parts by means of the Heitler–London relation. As examples are taken the linear conversions in the systems

$$H + H_2^{para} \rightarrow H_2{}^{ortho} + H, \quad H + HBr \rightarrow H_2 + Br, \quad H + Br_2 \rightarrow HBr + Br.$$

The course of a conversion is represented as a motion of a representative point over a surface which represents the binding energy of the atomic system, plotted in space as a function of the interatomic distances.

THE idea which F. London[1] has given of the possible course of a chemical reaction is that of an adiabatic interaction in which the energy of the system varies as a continuous function of the interatomic distances. He has also explicitly formulated these energy functions for the simplest case, in which reaction involves the reorganization of three or four monovalent atoms, for example,

$$Y + XZ \rightarrow YX + Z \tag{1}$$

or

$$VX + YZ \rightarrow VY + XZ. \tag{2}$$

In such cases the energy of the system can, according to London, be expressed, for any spatial configuration of atoms, as a function of the energy of the diatomic molecules which can be formed by combination in pairs of the reacting atoms. If, for example, two molecules of HI react,

$$2HI \rightarrow I_2 + H_2, \tag{3}$$

then in each stage of the conversion the energy can be expressed as a function of six interatomic distances provided that the energies of the single molecules HI, H_2 and I_2 are known as a function of the internuclear distances. In order to use London's formulae (which themselves are approximations) more accurately the fraction of the energy which is valence energy proper ("resonance energy") must be known and hence, by subtraction, the fraction

---

[1] F. London, *Probleme der moderen Physik* (Sommerfeld Festschrift), S. Hirtel, Leipzig, 1928, p. 104; *Z. Elektrochem.* **35**, 552 (1929).

which is "coulombic energy".[2] The expressions for the energy dealt with here are as follows:

For reactions of type 1 any arbitrary configuration of three atoms X, Y, Z is fixed by the distances a, b, c (Fig. 4.1). One can further designate as $\epsilon(a)$, $\epsilon(b)$, $\epsilon(c)$, respectively, the energy values

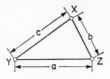

FIG. 4.1

of the isolated molecules ZY, XZ, XY for the internuclear distances a, b, c. Then the energy of each molecule can be expressed as a sum of the coulombic term $A(a)$, $B(b)$, $C(c)$ and the valence energy $\alpha(a)$, $\beta(b)$, $\gamma(c)$, respectively:

$$\epsilon(a) = A(a) + \alpha(a),$$
$$\epsilon(b) = B(b) + \beta(b),$$
$$\epsilon(c) = C(c) + \gamma(c).$$

The energy W_{abc} of the system can also be expressed by the equation

$$-W_{abc} = A + B + C + \sqrt{(\alpha^2 + \beta^2 + \gamma^2 - \alpha\beta - \beta\gamma - \gamma\alpha)}. \quad (4)$$

In the case of a reaction of type 2, in order to determine an arbitrary spatial configuration of four atoms, the values for six distances of the pairs of atoms VX, YZ, XZ, VY, VZ, XY must be used; they are designated in Fig. 4.2 by a_1, a_2, b_1, b_2, c_1, c_2.

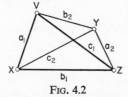

FIG. 4.2

---

[2] In what follows we do not take into consideration the zero-point energy; a correction for this will be introduced in the next section.

Here again the energy is a function of the energies of the isolated diatomic molecules, obtained by combining into pairs the reacting atoms. Again, besides the six valence terms $a_1(a_1)$, $a_2(a_2)$, $\beta_1(b_1)$, $\beta_2(b_2)$, $\gamma_1(c_1)$, $\gamma_2(c_2)$ there appear the coulombic terms $A_1(a_1)$, $A_2(a_2)$, $B_1(b_1)$, $B_2(b_2)$, $C_1(c_1)$, $C_2(c_2)$, and these combine together into the energy expression

$$W^{a_1 a_2 b_1 b_2 c_1 c_2} = A_1 + A_2 + B_1 + B_2 + C_1 + C_2 + [(a_1 + a_2)^2 + (\beta_1 + \beta_2)^2 + (\gamma_1 + \gamma_2)^2 - (a_1 + a_2)(\beta_1 + \beta_2) - (\beta_1 + \beta_2)(\gamma_1 + \gamma_2) - (\gamma_1 + \gamma_2)(a_1 + a_2)]^{1/2}. \tag{4a}$$

If the energy of a system as a function of the interatomic distances is known it is possible in principle to determine the activation energy of any reaction. The chemical initial and final states are two minima in the energy, and are separated by a series of energy hills. The lowest passage over these energy hills, which are responsible for the inertia of the reaction, gives the magnitude of the activation energy.

The activation energy may therefore be determined in principle by trying all possible ways which lead from the energy valley of the initial state to the energy valley of the final state, and establishing the value of the energy along the path in question. Of all the possible ways, the actual reaction path will be that which crosses the energy hills at the lowest height, and the activation energy will be given by the height of this energy ordinate.

At first sight such a survey of all possible passages seems too cumbersome. In the case of four atoms one would have to vary six coordinates in all possible ways, and even in the simplest case when only three particles are involved in reaction one has to represent the energy as a function of three coordinates and hence use four-dimensional space. It is therefore of great importance for the practical application of the theory that there are cases in which, on the basis of general considerations, a region can be defined where the passage of lowest energy should be sought. For the reaction of three atoms London succeeded in proving that such reactions take place with minimum activation energy when the three atoms approach along a straight line (see Fig. 4.3).

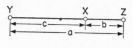

FIG. 4.3

For a reaction involving only three atoms, therefore, the number of independent variables on which the energy depends is reduced to two, because on the path of minimum activation energy (Fig. 4.3) $b+c=a$. In this case a realistic representation of the energy can be given by plotting it as a function of b and c. Proceeding in this way spatial energy representations were constructed for three reactions in which one hydrogen atom undergoes exchange with a different diatomic molecule according to the scheme $H + XZ \rightarrow HX + Z$. The cases $XZ = H_2$, $XZ = HBr$, $XZ = Br_2$ were selected, as follows

$$H + H_2^{para} \rightarrow H_2^{ortho} + H, \tag{5a}$$

$$H + HBr \rightarrow H_2 + Br, \tag{5b}$$

$$H + Br_2 \rightarrow HBr + Br. \tag{5c}$$

The working out of such a sequence of similar reactions which are at the same time widely different with respect to the effective energies and masses has, basically, the objective of deepening our insight into the mechanisms of adiabatic reactions. The question as to whether these particular reactions, and chemical reactions in general, do actually proceed adiabatically will not be discussed here. The fact that we find the adiabatic concept to be applicable to the extremely fast processes of the three examples given here (as well as to the high reactivity of free atoms and radicals in general), even after this more detailed analysis, certainly contributes to its plausibility; however the evidence is insufficient for this to be regarded as quantitatively proved.

Moreover, the significance ascribed to the adiabatic mechanism is accentuated because we can see no satisfactory representation of chemical reactions in terms of the occasionally discussed quantized course of chemical reactions.

The first attempt of this kind, by Villars,[3] is based on an assumption that would lead to an arrangement of reacting molecules such that no reaction would be possible at all. However, the reasonable considerations which lead J. Franck and E. Rabinowitsch[4] to describe a chemical reaction as a "radiationless transition" appear to us to provide, when developed quantitatively, nothing less than a proof of the important role of the adiabatic energy mechanism.

If we consider the best known bimolecular reaction $H_2 + I_2 \rightarrow 2HI$, and assume in the most favourable case that the total activation energy of about 40 kcal is in the form of vibrations of the hydrogen molecule, resulting in the expansion of the bond, then the distance between the two hydrogen atoms will reach as high a value as $1 \cdot 3$ Å.

It is not the case in any molecular lattice, however, that the positions of two atoms belonging to two different molecules lie so close together. Hence, if the hydrogen atoms of two activated HI molecules approach each other so closely in a collision which leads to the formation of $H_2 + I_2$, it follows that the activation energy not only extends the bonds but also acts effectively in reducing the repulsion forces. This fact, that the extension of a homopolar molecule lowers the repulsion barrier which separates it from neighbouring particles, is an essential result of the London theory and one which up to now has not been provided by any other theory. The fact that the effect of the activation consists, to a considerable extent, in the decrease of the repulsive forces therefore favours this theory as does the exceptionally high reactivity of free atoms and radicals for which up to now there has been no explanation other than that of an adiabatic mechanism. This is another argument for the significance of this theory.

[3] Villars, *Physik. Rev.* **34**, 1063 (1929).
[4] J. Franck and E. Rabinowitsch, *Z. Elektrochem.* **36**, 794 (1930).

The Components of the Activation Energy

In the work previously mentioned London attempted to simplify expression (4) for the energy to a form which only contains experimentally obtainable quantities. The simplification introduced by him was of a twofold nature. In the first place, he set the coulombic terms A, B, C equal to zero, and in the second place he also omitted the exchange term α which relates to the interaction of the two outer atoms. The result was that the height W_β of the energy barrier is reached at the point where $\gamma = \beta/2$, and the activation energy is determined by the difference between the valence energy β_0 of the XZ molecule in the initial state and the energy height W_β, giving $|\beta_0| - |W_\beta| = 0 \cdot 13$.

The quantity computed in this way for the activation energy may be called the "β-activation energy". It will be evident later, however, that besides this "β-component" there are important contributions to the activation energy resulting from the α-term and the coulombic term, previously neglected. The first tends to increase the activation energy, and therefore may be called the "α-component" or the "α-activation energy". The latter on the other hand tends to decrease the activation energy and hence may be called the coulombic deduction from the activation energy.

To these three components it is necessary to add a fourth; the zero-point energy also varies during this reaction and it is therefore necessary in determining the activation energy to take into consideration all maxima in the total energy, which is the sum of the potential energy and the zero-point energy. We will see later that the activation energy is always reduced as a result of this fourth term, but that the amount of reduction is in most cases small.

Further Organization of the Computations

In order to represent the energy surface for the linear conversions (Fig. 4.3) we must therefore know the dependence of the resonance energy and of the coulombic energy on the three interatomic distances b, c and $b+c=a$ and then, on the basis of the form of the expression

$$W'_{abc} = A(b + c) + B(b) + C(c) \tag{4'}$$
$$+ \sqrt{a(b+c)^2 + \beta(b)^2 + \gamma(c)^2 - a(b+c)\gamma(c) - a(b+c)\beta(b) - \beta(b)\gamma(c)},$$

we must examine the variation of the zero-point energy over this surface and add it, at every point, to W'_{abc}. Unfortunately we have only a very meagre knowledge of the functions A, B, C and of a, β, γ; we know them only for the H–H atom pair and even then only to a rather rough approximation. However, in order to proceed in spite of this, we assume as a first step that the total binding energy of the atom pairs in question consists of resonance energy. The activation energy which we calculate in this way will contain only the β- and a-components; it will lack the coulombic part and also the deduction due to the zero-point energy. The β- and the a-components will be too high owing to the neglect of the coulombic term and therefore our result will be, in some respects, too high.

The second step, in order for the simplest correction to be made first, should be the treatment of the zero-point energy; then, as a third step, a new representation should be made of the reaction of three H atoms on the basis of the coulombic and resonance energy curves calculated theoretically by Heitler and London. Fourthly a correction should be attempted, on the basis of results obtained here, for the activation energy computed in the first and second steps. Fifthly and lastly the dynamics of the conversion should be discussed, that is, the interplay of forces and impulses during the adiabatic conversion, and especially the question of the extents to which the kinetic and vibrational energies contribute to the heat of activation.

Energy Surfaces in the First Approximation

(a) *Energy Curves for the Atom Pairs*

The approximation which we wish to employ initially is, as just mentioned, that we consider the total binding energy as a resonance energy. If the binding energies of the three atom pairs in question (Fig. 4.3) are designated as $m(b)$, $n(c)$ and $l(b+c)$, then according to (4′) the energy surface will be

$$W''_{abc} = \sqrt{ \{ l(b+c)^2 + m(b)^2 + n(c)^2 - l(b+c)m(b) } $$
$$ - l(b+c)n(c) - m(b)n(c) \} . \qquad (4'')$$

Following a suggestion of H. Eyring[5] the functions l, m, n may be determined from the band spectra and the use of the approximate equation of P. M. Morse,[6] which gives the binding energy of a diatomic molecule as:

$$\epsilon = D(e^{-2k(r - r_0)} - 2e^{-k(r - r_0)}). \qquad (6)$$

The dissociation energy D (equal to the sum of the dissociation heat and the zero-point energy) is known for the diatomic molecules (H_2, HBr and Br_2) in question. Also the values of k can be determined, using the Morse procedure, from the known anharmonicity constant b'',

$$k = 0 \cdot 2454 \sqrt{(Mb'')}, \qquad (7)$$

where M is half the harmonic mean of the atomic weights, $M_1 M_2/(M_1 + M_2)$. Finally the normal internuclear distances, designated as r_0, of the diatomic molecules in question are known, so that equation (6) insofar as it is accurate makes it possible to determine the binding energy as a function of the internuclear distance (r).

For the three reactions selected for computations,

$$H + H_2^{para} \rightarrow H_2^{ortho} + H, \qquad (5a)$$

$$H + HBr \rightarrow H_2 + Br, \qquad (5b)$$

$$H + Br_2 \rightarrow HBr + Br \qquad (5c)$$

we require only three energy curves, namely those of the atom pairs H–H, H–Br and Br–Br (since the difference between the para- and ortho-hydrogen can be neglected). These curves, which

[5] H. Eyring, *Naturw.* **18**, 915 (1930).
[6] P. M. Morse, *Physic. Rev.* **34**, 57 (1929).

are given in Fig. 4.4, have been computed from equation (6) using the constants given in Table 1.

TABLE 1

Bond	U	v	b''	M	r_0	η	D	k
	(wave numbers, cm⁻¹)				(Å)	(wave numbers, cm⁻¹)		
H–H	37,000	4264	144·4	0·5	0·75	2132	39,132	2·09
H–Br	29,730	2559	—	0·988	1·42	1279	31,009	1·768
Br–Br	15,910	326·1	1·2	40·0	2·26	163	16·073	2·00

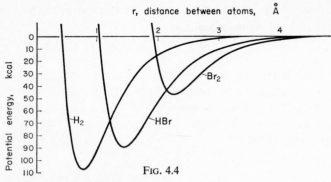

Fig. 4.4

In this table are entered:

$U =$ total dissociation energy,

$v =$ fundamental frequency,

$b'' =$ anharmonicity constant,

$M =$ half of the harmonic mean of the molecular weights,

$r_0 =$ normal internuclear distance,

$\eta =$ zero-point energy,

$D =$ potential energy of the bond,

$k =$ the constant of equation (6).

The constant k is computed for H_2 from b'' according to equation (7); for HBr and Br_2 from v and D according to the formula

$$k = 0 \cdot 2454 \, v \, \sqrt{\frac{M}{4D}} \,. \qquad (8)$$

A direct computation from the anharmonicity constants has been rejected for the latter two cases, because b'' for HBr is not known at all, and for Br_2 it is not known sufficiently accurately.

A reference point for the estimate of the approximation leading to equation (6) may be obtained by comparing the two k-values for hydrogen, derived on the one hand from equation (7) and on the other from equation (8). The first value (given in the table) is $2 \cdot 09$, the other is $1 \cdot 98$. The difference is so small that it can be neglected. For Br_2, the value of k computed from equation (7) is $1 \cdot 7$.

(b) Construction of the Energy Surfaces in the First Approximation

In Fig. 4.5 we have represented on a small scale, and therefore over a wide range, the resonance–energy contours for the reaction $H + H_2 \rightarrow H_2 + H$. This contour diagram has been constructed

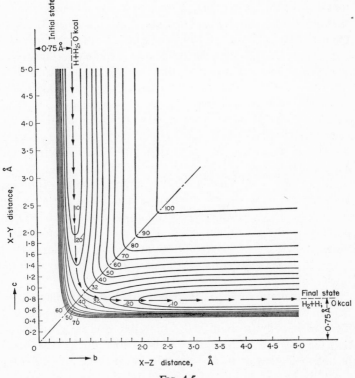

FIG. 4.5

using equation (4″), the coulombic terms being neglected. The distances b and c (compare Fig. 4.5a) vary here from 0 to 5 Å, which is considerably larger than the effective range of the valence forces. Therefore, on the right-hand side, we have practically

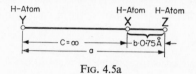

FIG. 4.5a

completely separated the H atom from the H_2 molecule and on the upper boundaries of the graph we have produced 3 H. The important domain for the exchange is reduced to a small region around the point $b = c = 1$ Å with a range of $0 \cdot 5$ to approximately 1 Å. In the next section this region will be considered with reference to an enlargement of the diagram, but now it is very important for us to elaborate on a wider domain of which this region is a part.

Since the energy difference between orthohydrogen and para-hydrogen has been neglected, in view of its small value, the final state in no way differs from the initial state, so that the diagram is symmetrical with respect to the diagonal at 45°, shown as a broken line. We select for the initial state that in which c is very large and $b = b_0 = 0 \cdot 75$ Å is the normal distance of the atoms in a hydrogen molecule and as the final state, that where conversely b is very large and $c = c_0 = 0 \cdot 75$ Å. Both these states lie at the bottom of the energy valleys, which extend to infinity at a distance of $0 \cdot 75$ Å parallel to the coordinate axes. The base of these valleys at infinity is selected as the zero of energy, so that the figures which are given on the contour lines of the diagram give the values of the function $D - W_{abc}''$.

The activation energy is therefore equal to the height of the barrier between the initial and final states. In the following it is assumed initially (without first testing the dynamic possibilities) that the reaction path consists of the path which passes as an orthogonal trajectory from the bottom of the initial valley over the top of the barrier (in Fig. 4.5 shown as a dashed line with arrows).

As can be seen, the highest point of this "reaction path" (the height of the barrier) is situated at $b = c = 0.95$ Å.

An energy wall, caused by the repulsive forces of the atoms, rises steeply between the "reaction path" and the coordinates. On the other side, towards the central region, the energy increases also, but slightly less steeply, since the work is here against the weaker attractive forces. Between these two mountain walls both valleys extend, rising gradually towards the diagonals, where they come together forming a saddle.

The extensive central plateau is the state of complete separation of the three atoms; the energy here attains the value D.

On the other side of the "reaction path", which corresponds to a compression, the energy must increase to infinity as the co-ordinates are approached. Here, however, an error in the approximation becomes apparent, because the rise takes place only to approximately 90% of D, after which there is a fall. This fall will cause more serious difficulties in later examples. We will have to assume that from the point where the decrease begins the diagram becomes invalid, and must be satisfied with that portion of the graph which lies outside this region.

The following explanation may be put forward for a further understanding of the provisionally introduced "reaction path", as well as a preparation for later applications of the diagram. A displacement of a point in the diagram parallel to a coordinate is solely a change of b or c; it therefore means that the distance between the central atom and one of the outer atoms has changed while the other remains constant. Furthermore each displacement that is slanted with respect to the coordinates indicates a simultaneous change of the distances b and c. If, therefore, a point in Fig. 4.5, coming from infinity, follows the "reaction path", as long as this path runs parallel to the c-coordinate, a hydrogen atom approaches the H_2 molecule in the direction of the molecular axis, without exerting any influence on this molecule. However, from the point where the "reaction path" deviates from a straight line, a further approach of that atom to the molecule causes an extension of the internuclear distance. When this "influence effect" becomes noticeable, the bottom of the valley starts to rise and with an increasing curvature the system moves up the valley until finally at the top of the saddle the activation energy is reached, both atom distances having become equal. This symmetrical transition state is shown in Fig. 4.6. The distances $b = c$ amount then to approximately $0 \cdot 95$ Å as compared with a normal distance of $0 \cdot 75$ Å in the H_2 molecule.[7]

In order for the energy height of the transition saddlepoint to be read off, the saddle region is represented in Fig. 4.7 on a larger scale. From it a value of about 30 kcal is obtained. It can be seen that allowance for the interaction of both outer atoms, which has its expression in the terms $l(b+c)$, produces a considerable contribution to the activation energy. The β-component, which would determine the activation energy if the α-term is neglected, has the value (corresponding to the expression $0 \cdot 13\ D$) of only 14 kcal. With this omission the reaction mechanism would be completely different; it would be obtained for the diagonal from eqn. (4) by neglecting $l(b+c)$ (since $b = c$):

$$W''_{abc} = m(b) = n(c).$$

Hence instead of a saddle there would be a basin whose minimum would be at the same height as the initial and final states and which would be separated from these states by two barriers of the height of the β-component. The position of these barriers, i.e. of their maxima, is determined, as already mentioned, by the relation $\beta = \dfrac{\gamma}{2}$; this relation is fulfilled in the case in question when the hydrogen atom has approached the H_2 molecule to a distance of $1 \cdot 34$ Å. Owing to the equality of the initial and final states the basin (replacing the previous saddle) will thus be bounded by a barrier at b equal to $1 \cdot 34$ Å and at c equal to $1 \cdot 34$ Å. If also the coulombic part of the bond energy is accounted for in a way that will be explained later, the basin between the barriers deepens by about 8 kcal so that the configuration H_3 must be completely stable at low temperatures.

[7] F. London, *loc. cit.*, p. 109.

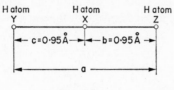

FIG. 4.6

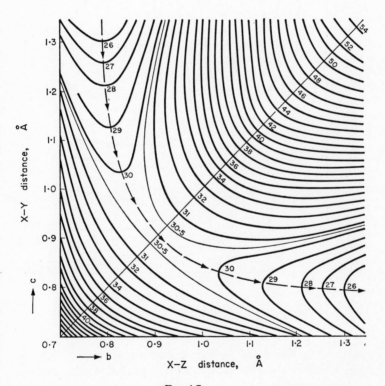

FIG. 4.7

Figure 4.8 is a part of the energy contour diagram for the reaction
H + HBr → H₂ + Br; this is (exactly as in Figs. 4.5 and 4.7)
computed from equation (4″) with the aid of the curves drawn in
Fig. 4.4. The significance of the distances b, c and a is moreover

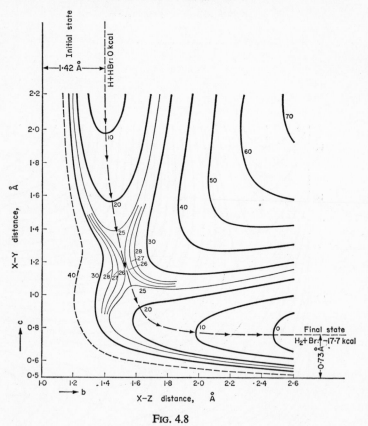

FIG. 4.8

particularly elucidated by the scheme shown in Fig. 4.8a. In
Fig. 4.8 the contour lines towards the compression side are drawn
as solid lines only up to the line corresponding to the unreal energy
decrease, mentioned already in the discussion of Fig. 4.5; the first

contour line at which this decrease is manifested is plotted as a
dotted line.

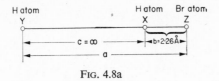

FIG. 4.8a

In contrast to the exchange of three H atoms we have in the
reaction $H + HBr \rightarrow H_2 + Br$ a genuine chemical reaction with
entirely different initial and final states. This is apparent from the
unsymmetrical shape of the diagram and further from the difference in levels of the initial and final states, corresponding to the
exothermicity of the reaction. It is especially worth noting that the
energy valley of the final state approaches so close to the c axis
that its end crosses the extended axis of the initial state valley; the
saddle, which separates both valleys, therefore lies nearly parallel
to the c axis.

Furthermore the "reaction path" passes over the activation
saddle without strong curvature and suffers a definite change of
direction only in the final-state valley. The dynamic importance of
this situation will be discussed later. Here it is to be noted only
that in the activated state, i.e. at the top of the saddle, the interatomic distance in the HBr molecule is only slightly extended; it
amounts only to about $0\cdot1$ Å above the normal distance. The
height of the saddle amounts to 26 kcal as compared to 11 kcal
which would be obtained for the β-component according to
$0\cdot13\ D_{HBr}$.

In the third example that we dealt with, namely the exchange
$H + Br_2 \rightarrow HBr + Br$, the distortion due to the unreal energy-
drop towards the compression side is so high that we can give in
Fig. 4.9 only outlines of the activation region. All the features
which were emphasized in the last example are even more en-
hanced here. The activation saddle lies on a linear extension of the
initial valley, so that the Br_2 molecule is not at all extended in the

c

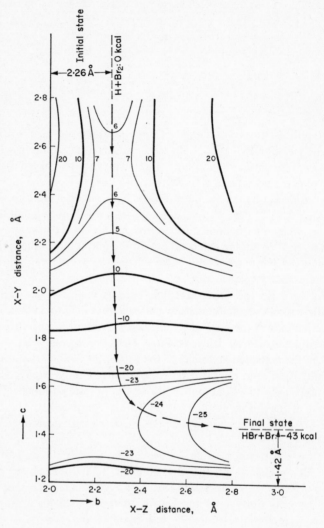

FIG. 4.9

transition state. This results from the fact that the total energy height of the activation saddle (6 kcal) originates from the "β-component" of the activation heat ($0 \cdot 13\ D_{Br_2} = 6$ kcal).

The Zero-point Energy Contribution to the Activation

The influence of the zero-point energy on the activation process can be seen if one follows the changes in the vibrations of the interacting atomic system during the chemical reaction. From Fig. 4.5 it can be seen that stationary oscillations, apart from the initial and final states, occur only at the top of the energy saddle. The zero-point energies in the initial and final states are

$$\eta_1 = \frac{h\nu_1}{2}, \quad \eta_2 = \frac{h\nu_2}{2} \tag{9}$$

and if, as in the exchange of three hydrogen atoms,

$$H + H_2 \rightarrow H_2 + H,$$

the final state is the same as the initial state $r_1 = r_2$ and $\eta_1 = \eta_2$, and the zero-point energy is

$$\eta_{H_2} = \tfrac{1}{2}\,h\nu_{H_2}.$$

In the case of Fig. 4.5, the vibrations taking place at the top of the saddle are directed along the diagonal; they correspond to symmetrical to-and-fro motions of both outer atoms with respect to the stationary central atom in Fig. 4.6. Since the curvature of the saddle, where the vibration takes place, is much flatter than the curvature of the bottom of the valleys for the initial and final states (Fig. 4.10) the quasi-elastic force which drives the outer atoms with respect to the central atom is much smaller than that which causes the extension of the atoms of a hydrogen molecule. Since the moving masses are the same in both cases it follows that the frequency of vibration is considerably smaller than the frequency of the hydrogen molecule. The zero-point energy at the top of the saddle, $\eta^* = h\nu/2$, is therefore much smaller than the initial value η_{H_2}. The total energy $W_{abc} - \eta$ therefore undergoes an *increase* resulting from the decrease in η during the course of the reaction; as a result of this, when the zero-point energy is taken into account the expression for the activation heat,

$$D - W^* - (\eta_0 - \eta^*) \tag{10}$$

(where D and η_0 refer to the initial state and W^* and η^* refer to the transition state) is smaller by this amount. If one considers that

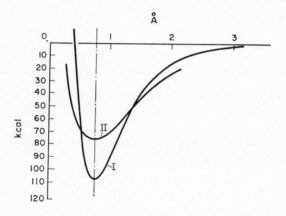

FIG. 4.10

the energy saddle with the inclusion of the coulombic terms will be even flatter (see Fig. 4.14), it follows that η^* can amount only to a small fraction of η. Hence in the computation of the activation energy of the reaction $H + H_2 \rightarrow H_2 + H$ almost the entire zero-point energy of the H_2, amounting to $6 \cdot 2$ kcal, should be subtracted. In the reactions of other molecules this subtraction must also be made, but it will be of small significance because the zero-point energy of molecules other than H_2 is very small.

New Computation for the Reaction $H + H_2 \rightarrow H_2 + H$ on the Basis of the Theoretically Determined Resonance and Coulombic Curves

In order to obtain an idea of the way in which the identification of the bond-energy with the resonance energy, which is the assumption of our calculations, influences the results we shall now discuss a new treatment of the one case in which the resonance

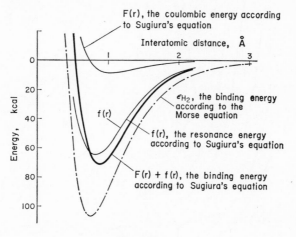

Fig. 4.11

and coulombic energies have been separately determined theoretically. The case we intend to discuss is the exchange $H + H_2 \rightarrow H_2 + H$, for the analysis of which only the functional relations of the resonance energy and the coulombic energy in terms of the interatomic distance H–H will be needed. These functions $f(r)$ and $F(r)$ have been determined on the basis of the equations of Sugiura,[8] which he developed from the theory of Heitler and London,[9] and are plotted in Fig. 4.11.[10] By analogy with

[8] Sugiura, *Z. Physik.* **45**, 484 (1927).
[9] Heitler and London, *Z. Physik.* **44**, 455 (1927).
[10] In Sugiura's paper $F(r) = E_1/(1 + \emptyset)$ and $f(r) = E_2/(1 + \emptyset)$.

Fig. 4.7, the resonance component remaining after the omission of the coulombic terms,

$$W_r = \sqrt{\{f(b+c)^2 + f(b)^2 + f(c)^2 - f(b+c).f(c) \qquad (11)}$$
$$\qquad -f(b+c).f(b)-f(b).f(c)\},$$

is given in the activation-saddle region in Fig. 4.12. It can be seen that the form of the energy map is slightly different from that of

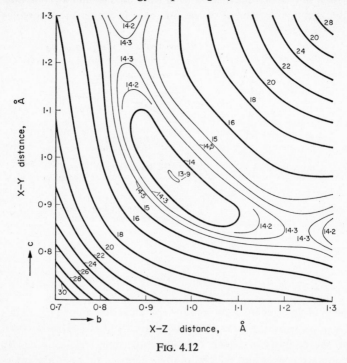

FIG. 4.12

Fig. 4.7, in that instead of a saddle a somewhat depressed plateau appears; this is separated from the initial and final valleys by two barriers, gradually rising from the high plateau. Such plateaux can, however, hardly occur in reality, because when the coulombic terms are added an energy basin appears instead of the plateau; this hole would be sufficiently deep to allow the formation of a stable H_3 molecule. We do not wish, however, to elaborate on this point, because the functions assumed are approximations that are still too rough. It is best seen how far they deviate from reality from a comparison of the curve for the sum $F(r) + f(r)$, which

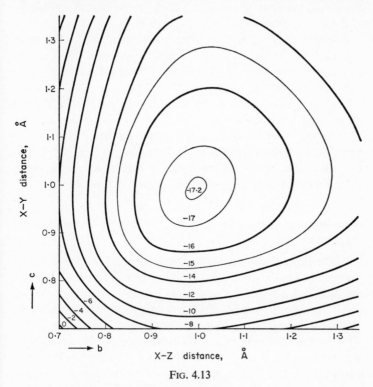

FIG. 4.13

indeed represents the total potential energy, with an empirically determined curve ϵ_{H_2}.

We prefer (in the approximation which is given in the next section) to rely on the function $F(r)$, whose absolute value is so low that the uncertainty in it is only of minor significance. The coulombic component

$$W_c = F(b+c) + F(b) + F(c) \qquad (12)$$

in the neighbourhood of the activation saddle is represented in Fig. 4.13. The coulombic surface has the shape of a funnel resulting from the additive character of this potential and the negative sign (in comparison with the activation energy). The level of the initial and the final states is taken as the zero-point as in Figs. 4.5 and 4.7.

Logically the next step would be to superimpose this coulombic potential well of Fig. 4.13 on the energy diagram of Fig. 4.12, and to construct from the two together the final form of the energy contours for the reaction $H + H_2 \rightarrow H_2 + H$. However, as already indicated, this operation does not seem to be justified on account of accumulated inaccuracies contributed by the approximations introduced. It seems to us that a better approach is the semi-empirical method, explained in the following section, which consists of a combination of the optically determined bond-energy curve and the theoretically-derived coulombic function.

Semi-empirical Method for the Final Treatment of the Reaction $H + H_2 \rightarrow H_2 + H$

It is assumed therefore that the binding energy, as postulated by Heitler and London, is composed of two types of energy, in such a way that the individual amounts of one type are superimposed additively on one another, while the components of the other kind are combining according to the mechanism of composition that is valid for the resonance terms. We further assume that the superimposable part is expressed by the function $F(r)$ given in Fig. 4.11, with about 10 to 20 % accuracy. The question is in which way this function, as part of the bond energy, modifies the picture which we obtained in Figs. 4.5 and 4.7 when we treated the bond energy entirely as the resonance energy.

The first step in correcting that picture is to correct the resonance barrier, upon which the coulombic basin given in Fig. 4.13 is then superimposed in order to obtain the final result. For the sake of consistency the function

$$\epsilon_{H_2}^* - F(r)$$

should be substituted in equation (4″). It did not, however, appear to us to be worth while to obtain this insignificant correction in such a cumbersome way, since the whole method is in any case fairly unreliable. We have therefore allowed for the decrease in ϵ_{H_2} due to the coulombic component by subtracting an amount which has its maximum value in the region of the activation saddle and diminishes towards the initial and final valleys, corresponding to the behaviour of the respective curves (Fig. 4.11). The amount subtracted is nearly constant in the region of the activation saddle and amounts to 8%; at the bottom of the valleys the amount subtracted is 1 kcal.

This correction affects the resonance energy contours in the sense that the value $|D| - |W''_{abc}| 0 \cdot 92 - 1$ appears instead of $|D| - |W''_{abc}|$ in the region of the activation saddle in which we are interested. The decrease in the resonance energies, as one can see from equation (4″), enters linearly into the expression for the exchange energy of the atomic system; in addition it is necessary, in order to correct the zero point, to subtract 1 kcal (the value of the coulombic energy at the bottom of the valley at infinity) from all of the values.

The second step in accounting for the coulombic energy terms consists in a superposition of the coulombic basin (Fig. 4.13) on the corrected resonance-energy contours. The result is seen in Fig. 4.14. One finds that the original saddle is greatly flattened and in one place there is even left a weak dent in the form of a high plateau. The height of this activation plateau amounts to about 19 kcal. The activation energy is obtained from this amount after correction for the zero-point energy. This correction is nearly equal to the total zero-point energy (see Fig. 4.10), making the activation energy about 13 kcal. According to the experiments of A. Farkas[11] the actual value is between 4 and 11 kcal.

The discussion of the two other examples

$$H + HBr \rightarrow H_2 + Br \quad \text{and} \quad H + Br_2 \rightarrow HBr + Br,$$

whose "resonance contours" we have computed (Figs. 4.8 and

[11] A. Farkas, *Z. Physikal. Ch.* (B) **10**, 419 (1930).

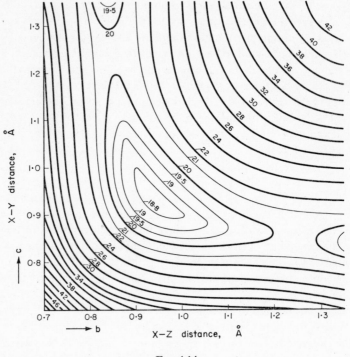

FIG. 4.14

4.9), can be done only very roughly beyond this first step. We assume that the depth of the coulombic basin is approximately the same as in the case of the three H atoms, so that after subtraction of the zero-point energy there remains in the first example an activation heat of about 10 kcal while in the second example the activation heat disappears completely.

Accurate measurements of the rate of the exchange in question are not available. However it is known from the experiments of Bonhoeffer[12] that atomic hydrogen reacts extremely fast with

[12] Bonhoeffer, *Z. physikal. Ch.* **119**, 385 (1926).

HBr as well as with Br_2, even at low pressures. Since this reaction may take place on the walls it should be noted that M. Bodenstein[13] concludes from his measurements of the rate of HBr formation that the two reactions $H + HBr \rightarrow H_2 + Br$ and $H + Br_2 \rightarrow HBr + Br$ have the same activation heats and that, on the other hand, it follows from the inhibition by Br_2 of the hydrogen–chlorine gas reaction[14] that the latter reaction must have a very small activation heat. Also it is known that similar reactions, involving chlorine instead of bromine, have a low inertia.

Thus, from the various examples it may be seen that the adiabatic mechanism interprets correctly one general feature of the reactions of free atoms, namely their relatively low heats of activation.

[The remainder of the paper, not reproduced here, is concerned with the dynamics of the motion of systems over potential-energy surfaces.]

---

13 M. Bodenstein, *Z. physikal. Ch.* **121**, 127 (1926).
14 St. v. Bogdandy and M. Polanyi, *Z. Elektrochem.* **33**, 554 (1927).

THE ABSOLUTE RATE OF REACTIONS IN CONDENSED PHASES†

W. F. K. WYNNE-JONES and HENRY EYRING

[With the development of statistical mechanics and the formulation of potential-energy surfaces the way was prepared for the development of an improved treatment of the absolute rates of reactions. A simple formulation that has been of very great value was made in 1935 by Henry Eyring (cf. also Paper 4). The paper reproduced below is not his first on this subject but is somewhat more comprehensive in the treatment of the activated complex than the earlier paper referred to in the introduction. In particular, the nature of the quasi-equilibrium between reactants and the activated complex is described at some length. Also, the theory is shown to explain the rates of certain reactions which on the basis of collision theory are abnormally fast or abnormally slow.

William Francis Kenrick Wynne-Jones was born in 1903 and was Leverhulme Research Fellow at Princeton University when he collaborated with Henry Eyring in the writing of the paper here reproduced. He was for a time head of the chemistry division of the Royal Aircraft Establishment, Farnborough, and at present is Professor and Head of the School of Chemistry at the University of Newcastle. In 1965 he was created Baron Wynne-Jones of Abergele.]

Abstract

The theory of absolute reaction rates is developed for condensed phases. The equation for the rate of a reaction of any order in any phase where the slow process is the passage over an energy barrier consists of the product of a transmission coefficient κ, a frequency kT/h, an equilibrium constant between an activated complex and the reactants and an activity coefficient factor. Previous theories of reaction rates such as Brönsted's, the collision theory of McC. Lewis, etc., are seen to be special cases of the general theory. A variety of examples are considered.

---

† A paper which appeared in *J. Chemical Physics* **3**, 492 (1935).

IN A previous paper one of us[1] outlined the theory of the absolute rate of reactions in terms of the activated complex and the probability of its formation from the reacting substances. Here we propose to consider in greater detail the properties of the activated complex and, by application of the equation already derived, to show how the rates of reaction in solution and at interfaces may be calculated. The relation of this treatment to certain empirical generalizations will be considered and the existence of numerous anomalous reaction rates will be explained.

Reaction rates in solution have usually been considered as too complex to yield to any simple treatment and the work of Christiansen[2] and of Norrish and Smith[3] seemed to show that rates in solution were abnormally slow. Moelwyn-Hughes[4] from an examination of a large mass of data showed that for many reactions the rate was not far from that calculated on the simple kinetic theory using a collision diameter of the same magnitude as for gaseous reactions. However, this attempt to explain reaction rates yields in general only the same qualified success that the method would give if applied to a calculation of equilibrium constants and, since we know that, for the latter, the method is quite crude, we must clearly seek for a more exact treatment of reaction rates.

In the reaction

$$n_1A_1 + \cdots + n_iA_i \to C \to m_1B_1 + \cdots + m_iB_i \tag{1}$$

nA and mB represent n moles of the reacting species A and m moles of the species B, respectively, while C represents the activated complex. The activated complex, C, differs markedly from an ordinary molecule only in that degree of freedom in which it is flying to pieces, that is to say, it has 4 instead of the usual 3 translational degrees of freedom. We can therefore write for the

[1] Eyring, *J. Chem. Phys.* **3**, 107 (1935).
[2] Christiansen, *Zeits. f. physik. Chemie.* **113**, 35 (1924).
[3] Norrish and Smith, *J. Chem. Soc.* 129 (1928).
[4] Moelwyn-Hughes, *Kinetics of Reactions in Solution*, Oxford Univ. Press.

equilibrium between the reacting molecules and the activated complex

$$K = K^{\ddagger} \frac{(2\pi m^{\ddagger}kT)^{\frac{1}{2}}}{h} = \frac{a^{\ddagger}}{a_1{}^{n_1} \cdots a_i{}^{n_i}} \frac{(2\pi m^{\ddagger}kT)^{\frac{1}{2}}}{h} \qquad (2)$$

where $a^{\ddagger}$ is the activity of the activated complex and a_1 to a_i the activities of the species A_1 to A_i. Later we consider the definitions of the standard state and introduce activity coefficients but first we shall consider states for which the activity may be identified with the concentration

$$K^{\ddagger} = c^{\ddagger}/(c_1{}^{n_1} \cdots c^{n_i}).$$

We now consider more closely the significance of K and $K^{\ddagger}$. The fact that the activated complex has a fourth translational degree of freedom does not affect the uniqueness of the expression we must write for the specific reaction rate but it does introduce an interesting arbitrariness into the expression for the concentration of the activated complex not present for molecules with but three degrees of freedom. Thus in (2) we have introduced K and $K^{\ddagger}$ both of which might be interpreted as equilibrium constants. For unit concentration of reactants K gives the concentration of activated complexes which would lie within a length of a cm along the fourth internal translation if for the whole cm the density in phase space were just that at the point of passage across the barrier. On the other hand $K^{\ddagger}$ is the equivalent expression for the concentration of activated complexes lying in a single unit, h, of the phase space along this internal translation instead of in the $(2\pi m^{\ddagger}kT)^{\frac{1}{2}}/h$ units for K. In (2) and throughout the paper we have chosen to define $K^{\ddagger}$ as our equilibrium constant. The advantage is that $K^{\ddagger}$ has the usual dimensions so that the specific rate of any reaction is simply obtained by multiplying by a transmission coefficient κ and the universal frequency kT/h. The rate equation is then:

$$\begin{aligned}
k' &= \kappa \cdot K \cdot \left(\frac{kT}{2\pi m^{\ddagger}}\right)^{\frac{1}{2}} \\
&= \kappa \cdot K^{\ddagger} \frac{(2\pi m^{\ddagger}kT)^{\frac{1}{2}}}{h} \cdot \left(\frac{kT}{2\pi m^{\ddagger}}\right)^{\frac{1}{2}} \\
&= \kappa \cdot K^{\ddagger} \cdot kT/h. \qquad (3)
\end{aligned}$$

Further substitution of

$$K^{\ddagger} = e^{-\Delta H^{\ddagger}/RT} e^{\Delta S^{\ddagger}/R}$$

gives

$$k' = \kappa \cdot e^{-\Delta H^{\ddagger}/RT} e^{\Delta S^{\ddagger}/R} kT/h. \tag{4}$$

These equations give the rate of a reaction in any phase provided that the slow process is the surmounting of a potential energy barrier. This is simply an alternative way of expressing equation (10) of the previous paper. We here employ the symbol k instead of c for the transmission factor in order to avoid confusion with concentration.

The equivalence of equations (3) and (4) to (10) of the previous paper is seen when we write the equilibrium constant $K^{\ddagger}$ in terms of the partition functions for the normal molecules and the activated complex

$$K^{\ddagger} = (F_a'/F_n)e^{-E_0/kT},$$

where F_a' is the partition function for all the degrees of freedom of the activated complex except the one along which decomposition is occurring, and F_n the partition function for all the reactants. Since the thermodynamical and statistical mechanical formulations are exactly equivalent, we shall be able, in our subsequent discussion, to use whichever equation happens to be the more convenient for our purpose. This is particularly valuable for reactions in solution as many of the properties of solutions have been experimentally determined and correlated on a thermodynamical basis although their interpretation by statistical mechanics has not often been possible except in rather general terms.

As previously indicated, whenever it is justifiable to regard the reactants simply as hard spheres, this treatment reduces exactly to the usual kinetic theory treatment, but we can show that such conditions will be rare and more or less accidental.

The Transmission Coefficient

The transmission coefficient, κ, represents the probability of the activated complex forming the products after crossing the barrier, i.e., not returning. If the activated complex makes on the average n crossings of the barrier before decomposition then $\kappa = 1/n$. From the shape of the potential surface for a given reaction it is possible to estimate the value of κ and frequently this value is very close to unity. The effect of the relative masses of the reacting atoms will be illustrated by consideration of the reaction of an atom with a diatomic molecule.

$$A_2 + A_1A_3 \rightarrow A_2A_1A_3 \rightarrow A_2A_1 + A_3.$$

Eyring and Polanyi[5] discussed this reaction in detail and pointed out that for the three atoms colliding along a straight line the mechanics of the reaction can be represented by a mass point moving on a surface made by plotting the potential vertically and using as horizontal axes the two distances between the atoms A_2 to A_1 and A_1 to A_3. These distances must be plotted not at right angles, however, but at an angle of $\theta = 90° - \sin^{-1}[(m_2m_3/(m_1 + m_2)(m_1+m_3)]^{\frac{1}{2}}$, where the m's are the masses of the three atoms. Thus we see that if either m_2 or m_3 is small compared with the other two masses, the acute angle, θ, approaches 90°. The surface itself consists of two confluent valleys parallel to the two axes. A mass point, representing our system, coming up one valley will, after passing over the intermediate low barrier, either be reflected and return or else pass into the second valley. In the latter case we have reaction. Which of these cases will arise depends on θ and therefore upon the relative masses. How much the two valleys round off into each other at their junction will also have its effect on κ and, as this rounding off will be the greater the more the coulombic or additive binding exceeds the exchange binding, κ should more nearly approach unity as the reactants become more metallic (since the coulombic energy then becomes notably great). Just how near κ approaches unity is thus seen to depend on the ratio of the colliding masses and the nature of the binding forces. Our present knowledge comes from consideration of the approximate surfaces available for some of the simpler reactions for which we can say that κ is of the order unity.

[5] Eyring and Polanyi, *Zeits. f. physik. Chemie* B **12**, 279 (1931).

On the other hand a type of reaction for which κ is often small is the adsorption of molecules on solid surfaces. Here the transmission coefficient is to be identified with the accommodation coefficient for that fraction of the molecules which strikes the surface with more than the necessary activation energy for adsorption.

Nature of Equilibrium in the Activated State

There is an important point regarding the activated complex which perhaps more than any other has been generally misunderstood. *It is not necessary that a large fraction of the activated complexes must fail to react in order that the equilibrium concentration be maintained. The activated state is not in general a state of indecision in which the activated complex is uncertain which way to proceed.* When κ is nearly unity an activated complex formed from the reactants will almost certainly decompose into products and whether it will do this or not is entirely independent of how fast the reaction is proceeding in the reverse direction. To say that activated complexes formed from reactants and from products collide and thus help to maintain equilibrium is without meaning as they are moving in spaces corresponding to different coordinates much as two trains passing on parallel tracks. By definition, for a gas reaction all molecules which are together at the activated state are included in the activated complex so that the direction in which the activated complex proceeds is a mechanically determined question and is not influenced by the number of activated complexes moving in the reverse direction. For solutions, the direction of decomposition is equally determined by mechanics if we include the solvent molecules in our activated complex. This is also true when we treat the effect of the solvent molecules as an external potential and integrate over the external coordinates of the reacting molecules to obtain K^+ in the way usual for equilibrium constants. The solvent molecules simply modify our potential surface but leave unchanged the concept of a mass point passing over a barrier. Now, since we can use equilibrium statistics when there are an equal number of activated complexes proceeding in both directions, we can still use it if those in one direction are completely suppressed, provided, of course, there is no autocatalysis. Thus if $\kappa = 1$ and we start out with reactants but no products the number of activated complexes going in the forward direction is exactly the equilibrium number and there are none going in the reverse direction. If some activated complexes collide with the surface and then return, the number moving in the forward direction will be fewer than the equilibrium number by the factor κ. If there is autocatalysis there are then two separate mechanisms which are important; one in which the activated complex contains the reaction products and one in which it does not. The previous considerations apply then to each separate mechanism whether there be one or a dozen. Chain reactions are simply built up of a series of successive reactions each of which is to be treated as we have indicated. This aspect of the theory of the activated complex is given particular prominence, as it is in sharp contrast with the commonly held view in which it is assumed that, only when most of the activated complexes fail to react (i.e., $\kappa \ll 1$), can one assume that the equilibrium numbers are moving in a given direction. The nature of the activated complex thus becomes clearly apparent when considered in terms of a potential surface.

Reaction in a Condensed Phase

As we have seen, we can write for the specific reaction rate constant

$$k' = \kappa \cdot K^{\ddagger} kT/h \cdots \tag{3}$$

or,
$$k' = \kappa \cdot (F_a/F_n)e^{-E_0/RT} kT/h \cdots, \tag{5}$$

providing our standard state has been defined so that the activities a_i can be identified with the concentration c_i. In the present unsatisfactory condition of the theory of condensed phases there is a real advantage in choosing the dilute gas as the standard state. Then for any condensed state we must write:

$$K^{\ddagger} = a^{\ddagger}/(a_1 \cdots a_n) = a^{\ddagger} c^{\ddagger}/(a_1 \cdots a_n \cdots c_1 \cdots c_n)$$

and for the rate $(c_1 \cdots c_n k')$

$$= \kappa \cdot c^{\ddagger} \cdot kT/h,$$

giving the specific rate

$$k' = \kappa \cdot K_0{}^{\ddagger} \cdot (a_1 \cdots a_n/a^{\ddagger}) \cdot kT/h \cdots, \tag{6}$$

where the a's are the activities and the activity coefficients a_i and a^+ approach unity as the gas becomes infinitely dilute. When the reaction proceeds in a phase other than the gaseous we still retain the gaseous state as the standard and as long as the gas phase is sufficiently near the ideal the a's have the significance of distribution coefficients between the two phases.

Our choice of the infinitely dilute gas as the standard state irrespective of the phase in which the reaction is proceeding has the great advantage that in the statistical mechanical calculation for the standard state we can restrict our attention to the reacting particles. If it is desirable to use statistical mechanics to calculate the activity coefficients, two avenues are open: we may treat the adjacent molecules either as entering into a larger activated complex, or as constituting a field in which the reacting molecules move. Either treatment simply has the effect of increasing the number of degrees of freedom for which the potential energy must be considered and leaves unchanged the general problem of the passage of an activated complex across a potential barrier.

For a reaction proceeding in solution we see that the rate is greater than in the dilute gas by the factor $(a_1 \cdots a_n)/a^+$ and it is clear that exact cancelation of the activity coefficients will be fortuitous and that, in general, a reaction is likely to proceed at a different rate in solution than in the gaseous state. For the exact calculation of the rate, a knowledge of distribution coefficients is essential and we will indicate briefly the method of introducing into our equation certain empirical generalizations.

For the distribution of a substance between its pure liquid phase and the gaseous phase several rules have been advanced of which the most valuable are Trouton's rule, its modification by Hildebrand and a more recent formulation by Langmuir.[6] Using

[6] Langmuir, *J. Am. Chem. Soc.* **54**, 2798 (1932).

Langmuir's equation (29) for the vapor pressure in atmospheres one readily obtains the equation $p = 2 \cdot 5 e^{3/2} T^{3/2} e^{-\Delta H/RT}$, which together with an average value of $0 \cdot 027$ for the concentration in moles per c.c. in the liquid phase, gives as a general equation for the distribution ratio between the liquid and gaseous states for nonassociated liquids

$$a = 5 \cdot 0 T^{\frac{1}{2}} e^{-\Delta H/RT}. \tag{7}$$

Over a temperature range extending from a few degrees absolute to some thousands of degrees, although a is of the order 10^3, this equation represents the data within 2 or 3-fold for such diverse substances as fused salts, the halogen acids and the elements both metallic and nonmetallic. If, instead of the value $0 \cdot 027$, the actual liquid concentration is employed, Langmuir's calculations show that the discrepancy is usually not more than 40 per cent. For certain associated liquids, however, such as water and ammonia, the numerical constant in the above equation is about 30 instead of 5.[7]

While these considerations apply to pure liquids we may readily extend them to solutions. If the solutions are ideal then clearly equation (7) applies; even if the solutions are nonideal, if the deviation arises principally from variation in the heat of vaporization then the equation is still applicable.

Frequently instead of using the approximate equation we may employ more accurate experimental data such as are available in the form of solubility coefficients of gases and vapor pressures.

Unimolecular Reactions

For any unimolecular reaction in the liquid phase we may write equation (6) in the form

$$k' = \kappa \cdot K_0^{+} a_1 / a^{+} \cdot kT/h \tag{8}$$

and we can distinguish three cases: (1) If the activated complex

[7] This equation applies only up to pressures of a few atmospheres. At the critical point a is of course unity. A more exact equation would take account of the decrease in the temperature dependence of a by including terms in the denominator of the general form $\{1 - \exp(-h\nu/kT)\}^{-1}$.

and the normal molecules both form ideal solutions in a solvent then, since by definition there is no heat of mixing, we see from equation (7) that $a/a^+ = 1$ hence

$$k' = \kappa \cdot K_0^{+} \cdot kT/h,$$

that is to say, the reaction will proceed at the same rate as in the gaseous state. (2) If the deviations from ideality are due mainly to changes in the heat of vaporization $(\varDelta H)$ then the a's may not cancel but their ratio will introduce a factor giving the equation

$$k' = \kappa \cdot K_0^{+} \cdot e^{-(\varDelta H - \varDelta H^+)/RT} \cdot kT/h$$

and the exponential term simply leads to a change in the apparent activation energy. (3) If the numerical factor is affected and to a different extent for the reacting molecules and the activated complex, as would happen if solvation of the various species occurred, we have the case where not only the apparent activation energy but also the factor multiplying it may be considerably altered.

If we know the reaction rates in solution and in the gas phase together with the partial pressures of the reactants we can deduce the partial pressure of the activated complex: this is true for reactions of any order. With the knowledge of the vapor pressure of the activated complex thus obtained we can go very far in excluding certain types of mechanism. Thus, if the data lead to a value of the vapor pressure of the activated complex which is compatible with equation (7) using a value of $\varDelta H$ which is at least approximately in agreement with Langmuir's[8] ideas of molecular surface energy, it is highly probable that the activated complex in solution is uncharged and homopolar. A change in mechanism as between the two phases will in general lead to an abnormal value for the apparent vapor pressure of the activated complex.

The classical example of a unimolecular reaction which has been studied in several phases by Daniels and his collaborators[9, 10] is the decomposition of N_2O_5 which in a variety of "inert" solvents proceeds with the same specific reaction rate as for the gas. By

[8] Langmuir, *Chem. Rev.* **6**, 451 (1929).

[9] Lueck, *J. Am. Chem. Soc.* **44**, 757 (1922).

[10] Eyring and Daniels, *J. Am. Chem. Soc.* **52**, 1473, 1486 (1930).

reference to the last column of Table V of the paper by Eyring and Daniels we see that a/a^+ for saturated solutions of N_2O_5 in nitromethane, carbon tetrachloride and liquid N_2O_4 are $1\cdot71, 2\cdot32$ and $2\cdot01$, respectively, whereas the next to the last column shows us that the corresponding values of a for nitromethane and carbon tetrachloride are 740 and 1108, respectively. These values of a are about what one would expect for an ideal solution. The values for the ratio a/a^+ indicate that the activated complex likewise has the vapor pressure of a practically normal molecule. N_2O_5 does not decompose in the interior of a crystal in spite of the fact that the activity of the normal molecule is the same as that in the gas so that the activity coefficient is approximately the same as in concentrated N_2O_5 solutions. The reason is that the activity coefficient of the activated complex is enormously high, since, for a molecule in a lattice to form the activated complex, it must displace a great number of molecules thus placing the complex itself under tremendous pressure and leading to a correspondingly high value for the activity coefficient, a^+.

Our reaction rate equation (3) in the form

$$k' = \kappa \cdot K \cdot (kT/2\pi m)^{\frac{1}{2}}$$

applies equally well for the process of sublimation, vaporization or desorption. When the activation energy is equal to the latent heat then K is the ratio of equilibrium concentrations of gas, c, to condensed molecules, c_s. The actual rate will then be given by

$$k'c_s = \kappa \cdot (K \cdot c_s)\, (kT/2\pi m)^{\frac{1}{2}}, \tag{9}$$

where the quantity $(K \cdot c_s) = c_1$ is simply the value deduced from vapor pressure measurements. The rate equation owes its simple form in this case to the fact that the activated complex is to be identified with the vapor molecules.

N_2O_5 decomposes at 45°C in nitric acid at $0\cdot042$ times and in propylene chloride at $0\cdot155$ times the rate in the gas; hence, we deduce that, for these solvents, either the vapor pressure of the normal molecules is low or the vapor pressure of the activated complex is high by comparison with the normal solvents. Eyring and Daniels' suggestion that the slow rate was due to the formation of a stable complex assumes, of course, that the low value of a/a^+ is due to a being low. Vapor pressure or solubility measurements will resolve this question. In general we should expect that abnormal solubility, which of course makes a low, will arise from abnormal solvation thus encasing the molecule in a sheath which may make a^+ high for the same sort of reasons as for a molecule in the interior of a crystal. Whether a^+ is increased or decreased depends on whether the solvent adds to the reactant in such a way as to strengthen or weaken the bond to be broken. There is a further point to be discussed in connection with this reaction. Since we know the actual rate of the reaction and also the heat of activation our theoretical equation enables us to evaluate the entropy of activation. The experimental results for the decomposition of N_2O_5 in the gas phase are represented by the equation

$$k' = 5 \cdot 08 \times 10^{13} e^{-24,700/RT}.$$

We take the transmission coefficient as unity, since it can hardly differ materially from this value for this reaction, and, by equating the above experimental result to the theoretical value for k', we find that the activated complex is $4 \cdot 27$ units richer in entropy than the normal molecules. This is a typical result for many unimolecular reactions.[11]

We will now use our general theory to calculate the steric factor for one of the elementary processes of a chain reaction postulated by F. O. Rice.[12] The reaction is the formation of a five-membered ring and a methyl radical from the normal hexyl radical: the activated complex will be an almost closed five-membered ring with the methyl radical in the act of leaving the ring. From the data of Parks and Huffman[13] we find that the entropy of cyclopentane is about 21 units less than that for normal pentane at 25°C; the formation of the ring changes 4 rotations into vibrations so that the difference in entropy should increase with temperature approximately as $4/2 \, R \ln T$. This leads to an entropy change for

[11] Polanyi and Wigner, *Zeits. f. physik. Chemie*, Haber Band, 439 (1928).

[12] F. O. Rice and K. K. Rice, *Aliphatic Free Radicals* (1935).

[13] Parks and Huffman, *Free Energies of Some Organic Compounds* (1932).

this ring closure at 650°C of about 25·6 units which corresponds to a steric factor of 2.5×10^{-6} as compared with Rice's estimate of 10^{-6}. This approximate calculation could be improved by taking account of the vibration frequencies in the ring by the formula $\prod_{i=1}^{4} 1/(1-e^{-hv_i/kT})$. However, as it stands, it is certainly correct to better than a power of ten. Thus, such a calculation gives a quantitative theory for the change of the steric factor with temperature, something quite impossible on previous ideas about reaction rates.

There is another type of unimolecular reaction occurring in solution which, although requiring an activation energy of as much as 35,000 calories, still proceeds at a measurable rate at room temperatures. In the past this has been interpreted on the assumption that the slow process is the transfer of energy in collision yielding an activated molecule with the energy, E, distributed among $F+1$ classical degrees of freedom giving the equation

$$k' = Ze^{-E/RT}(E/RT)^F 1/F!,$$

where Z is the number of collisions. This point of view forces Moelwyn-Hughes,[14] in order to explain the rapid reaction rate, to assume that the slow process is the transfer of energy between solvent molecules and the molecule decomposing. That, in solution, this should be the slow process, seems highly improbable when we realize that here if ever we should have the high pressure rate. That it is necessarily the mechanism is certainly untenable, as we see from our conception of an activated complex.

Although the theory invoking a collision and $F+1$ degrees of freedom is on the whole highly arbitrary it does at least take into account more degrees of freedom than are considered in the simple collision theory. Equation (4) shows that the effect of these other degrees of freedom manifested in the entropy of activation must be considered in chemical reactions.

An example of this type of reaction is the decomposition of triethylsulphonium bromide in various solvents. Taking the data

[14] Moelwyn-Hughes, *Kinetics of Reactions in Solution* (1933), p. 163.

of Corran,[15] which have been used by Soper[16] for demonstrating the analogy between the rate of the reaction and the entropy of the equilibrium process, we have calculated the entropies of activation which are given in column 3 of Table I.

TABLE I

DECOMPOSITION OF Et_3SBr IN BENZYL ALCOHOL-GLYCEROL MIXTURES

% Benzyl alcohol	ΔS	$\Delta S^{\ddagger}$
100	20·85	20·42
90·23	17·25	15·86
80·39	13·14	12·91
69·39	6·80	11·07

The striking correspondence between the entropy change for the reaction (column 2) and the entropy of activation shows us that the activated complex must resemble closely the products of the reaction. This is in marked contrast to the behavior of the N_2O_5 molecule where the activated complex had an entropy close to that of the decomposing molecule.

Bimolecular Reactions

When we come to bimolecular reactions in solution we find that there are reactions which according to the collision theory proceed too fast, too slow and at a normal rate.

Of the first type we have the reverse reaction to that considered above, the association of diethyl sulphide and ethyl bromide to form triethylsulphonium bromide. This reaction proceeds at a reasonable rate at 80°C in pure benzyl alcohol although it has the rather high activation energy of 28,400 calories as deduced from Corran's equilibrium measurements. A kinetic theory calculation yields for the collision diameter the value $1·14 \times 10^{-7}$ cm instead of the normal value of 2 or 3 Angstroms. Such an interpretation

[15] Corran, *Trans. Faraday Soc.* **23**, 605 (1927).
[16] Soper, *Chem. Soc.*, Discussion 45 (1931).

throws the entire responsibility on the translational coordinates and so leads to this impossibly large collision diameter. The conception of the activated complex places a part of the responsibility on the internal coordinates and we see at once that, since the unimolecular decomposition has an entropy of activation practically equal to the total entropy change for the equilibrium process, the bimolecular reaction must proceed with zero entropy of activation instead of the 7 units for a bimolecular reaction in solution[17] required to give agreement with kinetic theory.

A further example of an abnormally rapid reaction is the conversion of ammonium cyanate into urea

$$NH_4^+ + CNO^- = (NH_2)_2CO.$$

The energy of activation for this reaction in aqueous solution is 23,170 calories and from equation (4) we have $\Delta S^{\ddagger} = 0 \cdot 9$[18].

Moelwyn-Hughes[19] has compiled a list of reactions in solution which proceed at nearly the rate calculated from the collision theory. For reasons of space we will not include here the complete table of entropies of activation which we have computed for these reactions but merely state that where the reaction rate is just equal to that calculated by Moelwyn-Hughes the entropy of activation is 7 units. A departure by a factor of ten in either direction gives a corresponding change of $4 \cdot 6$ units in $\Delta S^{\ddagger}$. These "normal" reactions involve a neutral molecule and an ion as reactants and consequently the activated complex will carry the same charge as the ion and hence we should expect that the entropy due to solvation would not be much changed as between the reactants and the activated complex. It is, therefore, to be expected that such reactions should involve about the same entropy of activation as analogous gas reactions.[19a]

[17] For a bimolecular reaction at about 50–100°C a kinetic theory collision diameter of 2Å is equivalent to an entropy of activation of $6 \cdot 9$ while a diameter of 3Å corresponds to $5 \cdot 3$ entropy units.

[18] Walker and Hambly, *J. Chem. Soc.* **67**, 746 (1895).

[19] Moelwyn-Hughes, *Kinetics of Reactions in Solution* (1933), p. 79.

[19a] Even for a "normal" reaction the rate in solution is greater by the activity factor $a_1 a_2 / a^*$. Part of this factor is reflected in a smaller absolute value of the entropy of activation.

Of the reactions which proceed at an abnormally slow rate, one large class is that involving salt formation as in the Menschutkin reaction

$$(C_2H_5)_3N + C_2H_5I = (C_2H_5)_4NI.$$

A number of these reactions have been tabulated by Moelwyn-Hughes[20] and the values we have calculated for their entropies of activation range from 25 to 50 units. In general, if for a given

TABLE II

$(C_2H_5)_3N + C_2H_5Br$

IN ACETONE–BENZENE MIXTURES

% Acetone	$\Delta H^{\ddagger}$	$- \Delta S^{\ddagger}$
100	11,710	41·29
80	12,100	40·32
50	12,040	41·64
20	12,180	42·42
0	11,190	48·09

equilibrium there is a large entropy change, it is inevitable that either the forward or the reverse reaction proceeds at an abnormal rate. Not many equilibrium data are available but from the results of Essex and Gelormini[21] for the reaction

$$C_6H_5N(CH_3)_2 + CH_3I = C_6H_5N(CH_3)_3I$$

occurring in nitrobenzene at 60°C we calculate the entropy change for the equilibrium is 54·1 units while $-\Delta S^{\ddagger}$ is 37·5. This suggests that the activated complex is strongly polar and has many solvent molecules oriented around it. In a nonpolar solvent instead of orientation of solvent molecules there will be orientation of the reactants and products. This change in behavior on going from a polar to a nonpolar solvent is well exemplified by the data for the reaction between triethylamine and ethyl bromide in mixtures of acetone and benzene. We see from the figures in column 3 that the

[20] Moelwyn-Hughes, *Kinetics of Reactions in Solution* (1933), p. 111.
[21] Essex and Gelormini, *J. Am. Chem. Soc.* **48**, 882 (1926).

entropy of activation remains nearly constant so long as there is an appreciable number of acetone molecules which can be oriented around the activated complex but in pure benzene there is a markedly greater decrease in the entropy of the activated complex.

The known behavior of electrolytes in different solvents and in particular the investigations of Kraus and his co-workers[22] into the properties of electrolytic solutions in nonpolar solvents is in harmony with these suggestions of solvent orientation and solute association.

Quite recently Williams and Hinshelwood[23] have studied the benzoylation of certain amines and in conformity with the previous

TABLE III

BENZOYLATION OF AMINES IN BENZENE

Amine	Acid chloride	$\Delta H^{\ddagger}$	$-\Delta S^{\ddagger}$
H	p.NO_2	5900	40·22
p.CH_3	H	6800	38·54
H	p.Cl	7000	39·97
H	H	7350	39·62
H	p.CH_3	8000	39·24
p.Cl	H	7600	41·83
m.NO_2	H	10,500	39·17
p.NO_2	p.NO_2	10,400	42·52
p.NO_2	H	11,800	39·50

work of Bradfield et al.[24] have shown that while these reactions proceed at a rate about a million times slower than that predicted by kinetic theory considerations the steric factor that they introduce is nearly the same for each reaction. In Table III we give the values of $-\Delta S^{\ddagger}$ for these reactions.

The reactions all involve substituted anilines reacting with

[22] Kraus and Hooper, Proc. Nat. Acad. Sci. 19, 939 (1933); Kraus and Vingee, J. Am. Chem. Soc. 56, 511 (1934); Kraus, Trans. Electrochem. Soc. 66, 179 (1934).

[23] Williams and Hinshelwood, J. Chem. Soc. 1079 (1934).

[24] Bradfield, Jones and Spencer, J. Chem. Soc. 2907 (1931).

various substituted benzoyl chlorides and we indicate in column 1 the particular substituent in the aniline molecule and in column 2 the substituent in the benzoyl chloride. The reaction in each case is of the type

$$C_6H_5NH_2 + C_6H_5COCl = C_6H_5NHCOC_6H_5 + HCl$$

and the activated complex will consist of the two reacting molecules linked between the N atom and the carbonyl carbon atom with hydrogen and chlorine ions in the act of splitting off. This activated complex will be strongly polar and the high entropies of activation are in accord with this. It is interesting to note that, whereas the energies of activation vary considerably, the entropy is practically the same for each reaction. This constancy of the entropy of activation is to be expected since the strongly polar portion of the activated complex is the same in each case. On the other hand the variation of the energy of activation receives a simple explanation in terms of the dipole effects of the substituent groupings as has been pointed out by Williams and Hinshelwood who find that the equation of Nathan and Watson[25]

$$E = E_0 \pm c(\mu - a\mu^2)$$

is applicable for these reactions.

Baker and Nathan[26] have recently published the results of a study of the reaction between various substituted benzyl bromides and pyridine or α-picoline. The values of $-\Delta S^+$ calculated from their data vary from 20 to 40 units depending upon the particular reaction and the solvent. These values are similar to those quoted above for the Menschutkin reaction and support the idea that the activated complex has a strongly polar structure possibly of the type suggested by Baker and Nathan

$$C_5H_5N^+ArBr^-.$$

One more point should be emphasized in connection with reactions in which the charge on particular atoms changes during the reaction. If the change in charge occurs before or after the system

[25] Nathan and Watson, *J. Chem. Soc.* 2436 (1933).
[26] Baker and Nathan, *J. Chem. Soc.* 519 (1935).

reaches the activated state then our treatment of the slow process as adiabatic is complete. If the change in charge coincides with and is partly responsible for the slow process (i.e., the potential surface for our reacting system is formed from two surfaces corresponding to different atomic charges, intersecting at the activated state) then a part of the chemical inertia which we have included in the term ΔS^+ should be interpreted instead as a slowness of transition from one surface to the other. In the present state of our knowledge of potential surfaces for reaction in solution a more definitive statement can hardly be made. However, when there is a large overall entropy change for the reactions it is very reasonable to assume that the observed slowness arises simply from the low entropy of the activated state. The possibility that a very slow reaction may be nonadiabatic must not be overlooked. If the apparent entropy of the activated state is as high or higher than would be expected for a similar molecule the reaction is probably adiabatic otherwise it is suspect.

Acid and Base Catalysis

Another group of reactions of great interest is that included under the general description of acid and basic catalysis. These reactions have been intensively studied of recent years especially in the laboratories of Brönsted and Dawson and an excellent review of their main features is given in an article by Pedersen.[27] Brönsted and Pedersen first proposed the relation

$$k' = GK^x$$

where k' is the specific rate constant for a reaction which treats the catalyst as a reactant and where K is the dissociation constant of any acid or base. G and x are independent of the catalyst for a given reaction at a definite temperature. If x is independent of the temperature we have

$$d \ln k'/dT = x \cdot d \ln K/dT + d \ln G/dT \tag{10}$$

and, as ΔH for the ionization of many weak acids with a

[27] Pedersen, *Den Almindelige syre og Basekatalyse*, Copenhagen 1932.

dissociation constant of about 10^{-5} at room temperature is approximately zero,

$$d \ln k'/dT \sim d \ln G/dT,$$

that is to say the energy of activation is nearly the same for different catalysts. This is shown by the work of Kilpatrick, Pedersen and Smith.[28, 27] Without introducing this approximation we may write (10) in the form

$$(\Delta H^{\ddagger}/RT - \Delta S^{\ddagger}/R) = x \cdot (\Delta H/RT - \Delta S/R) - \ln G \qquad (11)$$

and set

$$\Delta H^{\ddagger} = x \cdot \Delta H - \Delta H_G, \qquad (11a)$$

$$\Delta S^{\ddagger} = x \cdot \Delta S + \Delta S_G, \qquad (11b)$$

where ΔH_G and ΔS_G are the energy and entropy components of $\ln G$. Pedersen has shown that in the bromination of acetoacetic ester the above equation (11a) for the energy of activation holds quite well using the same x for the various basic catalysts that he employed.

The results of Smith for the iodination of acetone enables us to calculate the values of $\Delta S^{\ddagger}$ given in Table IV. The first three columns are self-explanatory, the fourth contains the value of the

TABLE IV

IODINATION OF ACETONE IN AQUEOUS SOLUTION

Catalyst	$\Delta H^{\ddagger}$	$-\Delta S^{\ddagger}$	$-\Delta S$	$-(10 \cdot 4 + \Delta S/2)$	$P/K\frac{1}{2}$
$OH_3{}^+$	20,680	10·4	0	10·4	0·021
$Cl_2CH.COOH$	19,230	16·9	(13·0)	(16·9)	0·025
$ClCH_2.COOH$	19,230	21·2	25·1	23·0	0·016
$CH_3.COOH$	20,010	24·2	30·1	25·4	0·033
$C_2H_5.COOH$	19,370	26·7	30·9	26·3	0·011

entropies of ionization of the various acids calculated from the data of Harned and Embree.[29] To make the data consistent with

[28] Kilpatrick and Kilpatrick, *J. Am. Chem. Soc.* **53**, 3898 (1931); Smith, *J. Chem. Soc.* 1744 (1934).

[29] Harned and Embree, *J. Am. Chem. Soc.* **56**, 1050 (1934).

the assumed formula for hydrogen ion OH_3^+ the usual dissociation constants of all the other acids are divided by $55 \cdot 5$. In column 5 we give the values obtained by adding half the entropy of ionization to the entropy of activation for the hydrogen-catalyzed reaction. Columns 3 and 5 now show a very interesting agreement. Column 6 contains the figures given by Smith for the steric factor (which on the collision theory must be introduced to account for the observed rates) divided by the square root of the dissociation constant. The figures given for dichloroacetic acid in columns 4 and 5 are those required to fit the experimental entropy of activation given in column 3. In the absence of accurate data for the dissociation constant of this acid this value of ΔS may be regarded as a provisional estimate.

It is not necessarily true that x has the same value in equations (11a) and (11b) and it may well be that some of the departures from linearity in the plots of log k' *vs.* log K are due to this cause.

The Brönsted Equation

We have seen that equation (6) gives the rate of a reaction in any phase in terms of that in the gas phase. However, in any given solvent it is convenient to employ activity coefficients which are defined with respect to a standard state in that solvent—the infinitely dilute solution—and consequently we will separate the factor $(a_1 \cdots a_n)/a^+$ into two terms. The first term will represent the distribution coefficient between the standard states for the liquid and gas whereas the second will give the activity coefficient factor in the solvent. We then rewrite equation (6) in the form

$$k' = (\kappa \cdot K_0^+ \cdot ((\beta_1 \cdots \beta_n)/\beta^+) \ (kT/h)) \ (f_1 \cdots f_n)/f^+. \qquad (12)$$

The factor $(f_1 \cdots f_n)/f^+$ is simply the Brönsted activity factor and the remarkable success in the interpretation of salt effects which has been achieved by application of the Brönsted equation is too well known to require recapitulation here. The above equation goes beyond that of Brönsted in that it provides a theory for the quantity in parenthesis.

The critical complex of Brönsted's theory has been variously interpreted. Brönsted[30] himself refers to it as being formed by collisions between the reacting molecules and as decomposing instantaneously into the products, while Bjerrum[31] treats it as a molecule with a life which is long compared with the duration of a collision. As we have already stated there is no such ambiguity about the description of the activated complex which is simply a molecule in statistical equilibrium with the reactants and which is actually in the act of flying to pieces having a mean life which we can take as h/kT. The fleeting existence of the activated complex might be regarded as making inapplicable the usual rules for activity coefficients and of course for electrically charged complexes we have the problem that whereas the time of relaxation of the ion atmosphere is of the order 10^{-10} second the life of the activated complex is only $1 \cdot 6 \times 10^{-13}$ second at ordinary temperatures. However, this apparent difficulty is not real, since the molecules which come together and form the activated complex will be those which are already within one mean free path and will therefore have reciprocally modified their ionic atmospheres in such a way as to make a combined atmosphere approximating to that of the activated complex. The Brönsted activity factor therefore gives a precise description of the behavior of a reaction with changing medium and the activity coefficients involved are purely thermo-dynamical quantities. Of course, it is true that localization of charges in the activated complex will make its activity coefficient depend upon the ionic strength in some less simple manner than for symmetrical ions, but this problem is not peculiar to reaction rate studies but occurs equally in ordinary equilibria. La Mer and Kamner[32] have made accurate measurements on the rate of the reaction between β-bromopropionate and thiosulphate ions and have found that, on the mechanism they assumed, the activity factor is of opposite sign to that predicted from the Brönsted equation. They suggest as explanation that the localization of the

[30] Brönsted, Zeits. f. physik. Chemie 102, 109 (1922); 115, 337 (1925).
[31] Bjerrum, Zeits. f. physik. Chemie 108, 82 (1924).
[32] La Mer and Kamner, J. Am. Chem. Soc. 53, 2832 (1931).

D

charge on the bromopropionate ion results in an abnormal activity coefficient for the activated complex. Sturtevant[33] has shown that dissymmetry of the field does not affect the validity of the Debye-Hückel limiting equation for activity coefficients and, since we have shown that the same considerations are valid for the activated complex as for ordinary molecules, we conclude that, if the results of La Mer and Kamner are as accurate as they appear, some other mechanism must be sought for this reaction.

The present theory of rates reduces the problems of kinetics to a thermodynamic calculation of an equilibrium constant for the activated complex and the calculation of a transmission coefficient which is frequently very near unity. The principal problems are thus seen to be those always met in calculating an entropy and a heat of reaction. The difficulties in the simple kinetic picture are thus seen to disappear when viewed in the light of the general theory.

One of us (W. F. K. W-J) is indebted to the Leverhulme Trustees for a fellowship and to the University of Reading for leave of absence.

Note added in proof: Of the two definitions for the equilibrium constant used here and in the previous article, Evans and Polanyi (*Trans. Faraday Soc.* **21**, 875 (1935)), in a paper which has appeared while this article was in press have employed the one which requires multiplication by the velocity normal to the energy barrier (and of course the transmission coefficient κ) to give the absolute rate. We also have chosen this definition for processes (such as vaporization or adsorption) where the energy barrier is localized between two phases. For processes inside a single phase we use the definition for which the equilibrium constant must be multiplied by the frequency kT/h and the transmission coefficient κ to give the absolute rate. This difference in choice of the equilibrium constant is a formal one which will be ultimately decided on the basis of utility.

Our value of $5T^{\frac{1}{2}} = 87$ at room temperature (see equation (7)) for the apparent ratio of bimolecular collisions in the liquid to the gaseous state agrees with their factor of about 100.

[33] Sturtevant, *J. Chem. Phys.* **3**, 295 (1935).

PART III

Unimolecular Reactions

PAPER 6

THE RADIATION THEORY OF
CHEMICAL ACTION†

F. A. LINDEMANN

[At the time that this Faraday Society Discussion was held there was
considerable interest in the hypothesis that the activation of molecules
in thermal reactions is brought about by absorption of infra-red radiation
emitted by the walls of the vessel. This mechanism of activation had been
advocated by F. Perrin in the case of unimolecular reactions, for which
activation by collisions did not seem at first sight to offer a satisfactory
explanation, since the frequency of collisions must depend on the square
of the concentration. Lindemann, in the contribution that is reproduced
below, showed how it is possible under certain circumstances for second-
order activation by collisions to lead to first-order kinetics. One of the
consequences of his theory was that unimolecular reactions should
become second-order at sufficiently low pressures, and later work confirmed that this change of order does take place.

Frederick Alexander Lindemann (1886–1957) was for many years
Professor of Experimental Philosophy at Oxford. He did distinguished
work in many areas of physics, including astrophysics. He was created
Baron Cherwell of Oxford in 1941, and Viscount Cherwell in 1956; he was
personal assistant to the Prime Minister (Winston Churchill) in 1940,
and Paymaster-General in 1942–5 and 1951–3.]

PROFESSOR F. A. LINDEMANN: In view of the short time at my dis-
posal I will confine my remarks to the criticism of the radiation
theory of chemical reaction velocity, which I myself have published.

If Professor Perrin's premises are absolutely correct his argument in favour of the radiation theory is no doubt sound. Briefly,

† A contribution to a Faraday Society Discussion published in the
Transactions of the Faraday Society, **17**, 598 (1922).

he says it is experimentally established within certain limits that reaction velocity is not changed if the concentration, and therefore the number of collisions per second, is diminished. He assumes that he may extrapolate to such low concentrations that practically no collisions occur and that the reaction velocity will then still be constant. He concludes, therefore, that reaction velocity cannot be due to collisions but can only be due to some outside agency, which must obviously be radiation. The argument is convincing if it is true that reaction velocity is really independent of the number of collisions; but surely our evidence of this is quite inadequate at low pressures.

My difficulty may be recapitulated as follows: If reaction velocity is proportional to radiation density of a given wavelength, then the inversion of sucrose (a case studied by Professor Lewis, which has a temperature coefficient of $4 \cdot 13$ for $10°$) must be determined by the radiation density of wavelength $1 \cdot 05\mu$. The radiation density of this wavelength is 5×10^{18} times greater in sunlight than in the dark, yet the reaction proceeds at approximately the same rate.

Professor Perrin attempts, as I understand, to meet this by assuming that one has to deal not with the density of radiation ur of frequency ν, but with a number of absorption bands ν_1, ν_2 such that $\nu_1 + \nu_2 + \cdots = \nu$. But if the velocity is proportional to $u\nu_1 + u\nu_2 + \cdots$ we shall not find Arrhenius's well-known expression, since $\dfrac{1}{\Sigma u_v} \dfrac{d}{dT} \Sigma u_v$ is not proportional to A/T^2. In any case the assumption of five or more critical frequencies, and some such number is required, seems to me very artificial.

Professor Lewis endeavoured to meet my criticism in another way, which I had myself considered but rejected on account of certain results it implies. He said that in sunlight the radiation density of the critical frequency inside the liquid is scarcely increased, since practically all the radiation of that frequency is absorbed in the first millimetre. As a matter of fact one would have a somewhat greater depth, since Professor Lewis's temperature coefficient requires a wavelength of $1 \cdot 05\mu$. But if one takes

1 mm, surely in any ordinary beaker or test-tube an increase in the reaction velocity of 5×10^{13} in the first millimetre will be perceptible even if the next 100 mm are unaffected. If the solution were stirred or the test-tube shaken all the liquid would soon have come within 1 mm of the surface and the reaction would be completed. Professor Lewis also suggested that in sunlight one is dealing with directed radiation and not temperature radiation. Frankly, I cannot see how a molecule can differentiate between them. Suppose, say, one is dealing with a molecule capable of dissociating and believes dissociation to be caused by radiation. How is the molecule to know whether the radiation which falls on it is temperature radiation or not? A single molecule has no temperature and cannot possibly tell whether it is in equilibrium with the radiation. If the radiation density is the determining factor it cannot matter how this radiation density is brought about.

After all this criticism, I suppose I had better put forward a constructive suggestion intended to meet Professor Perrin's difficulty. Taking the simplest possible case, a dissociation, we may assume that dissociation can only take place if the centrifugal force due to rotation is above a certain value. There is nothing new in this assumption and it leads to Arrhenius's expression for the temperature coefficient. It leaves Professor Perrin's difficulty untouched, however, since without further assumptions it would follow that all molecules rotating fast would dissociate and the reaction velocity would depend upon the rate at which the Maxwell–Boltzmann distribution was re-established which depends upon the collision frequency or density. The difficulty vanishes, however, if we assume that a molecule even under a large centrifugal force does not dissociate instantly but on the average only after a time greater than that necessary to re-establish the Maxwell distribution. In this case one would effectively always have the Maxwell distribution and the number to break up per second would be proportional to $e^{-Q/RT}$, so that $\dfrac{1}{v}\dfrac{dv}{dt}$ would be Q/RT^2 as it should be.

The assumption of such an average life after the critical centrifugal force has begun to act seems fairly plausible. It is reasonably certain that molecules are held together, at any rate in part, by electrons in motion and it seems quite plausible to assume that the binding force depends upon the relative position and state of motion of the atoms and electrons. In this case there will obviously be certain positions and states of motion at which the binding force is a minimum and it is evident that dissociation will take place, given sufficient centrifugal force, when these occur. Hence the above-mentioned time between attaining a sufficient velocity of rotation and actual dissociation will be on the average half the time between the recurrence of these positions of minimum force. It should also presumably be proportional to the reduced absolute velocity of reaction.

If this view is true, of course there must be a limit beyond which the experimental law that reaction velocity is independent of concentration, ceases to hold, namely at concentrations at which the time between collisions is of the same order as the average life of the molecule after attaining its critical velocity. At very low pressures, especially in simple slow reactions with low temperature coefficients, such deviations might perhaps be observed and would form an interesting check on the theory. In liquids of course no such effect could occur, since in these the Maxwell distribution must be rapidly re-established by collisions with the solvent molecules whatever the concentration of the solute.

At present of course there is no more evidence in favour of this view than of any other, but it seems to me free from any insuperable objections and it does get over the initial dilemma which forced Professor Perrin as he has said to adopt the radiation hypothesis.

[It is to be noted that Lindemann emphasizes the role of rotational energy in leading to dissociation. Subsequent work has shown that the vibrational energy plays a more important role.]

ON THE THEORY OF UNIMOLECULAR REACTIONS†

C. N. HINSHELWOOD

[According to Lindemann's theory the second-order low-pressure rate constant of a unimolecular reaction is equal to the rate constant for the process of forming an activated molecule A* by the collision process A + A → A* + A. On the simplest view this rate constant would be $Ze^{-E/RT}$ where Z is the ordinary collision number. It was, however, found that the use of this formula led to a value for the rate of activation that was much lower than the experimental value. The pressure range over which the kinetics changes from first order to second order varies inversely with the rate of activation, and it was found that the use of the formula $Ze^{-E/RT}$ led to the prediction that the order should change at a much higher pressure than was actually found experimentally in a number of cases. Hinshelwood's contribution in this paper was to show that a much greater rate of activation can be calculated if one takes into account the fact that the energy of activation is distributed between n degrees of freedom in the molecule.

Cyril Norman Hinshelwood (b. 1897) was at the time of writing this paper Fellow and Tutor of Trinity College, Oxford; he was Dr. Lee's Professor of Chemistry from 1937 until his retirement in 1964.

He has made important contributions to many fundamental problems in chemical kinetics; particular reference may be made to his pioneering work on unimolecular gas reactions, the theory of chain reactions, reactions on surfaces and processes occurring in the bacterial cell. His many honours include the Presidency of the Royal Society (1955–60), knighthood (1948), a Nobel prize (1956) and the Order of Merit (1960).]

IT IS usually supposed that the energy of activation is confined to a few degrees of freedom. On this assumption the expression for the

† The first portion of an article which appeared in the *Proceedings of the Royal Society*, A, **113**, 320 (1927).

chance that a molecule possesses energy in excess of E is $e^{-E/RT}$ with sufficient exactness for most purposes.

The rate of bimolecular reactions is quite adequately expressed by the equation *number of molecules reacting* $= Ze^{-E/RT}$, where Z is the number entering into collision.

If molecules are activated by collision in unimolecular reactions and the mechanism proposed by Lindemann[1] operates, the energy E ought in unit time to be communicated to a number of molecules not merely equal to, but much greater than, the number reacting. This is because most of the activated molecules have to be assumed to lose their energy again. But in all known unimolecular reactions many more molecules undergo chemical change than the expression $Ze^{-E/RT}$ allows for. This has been considered to raise a serious difficulty, the only ways round which are assumptions about "reaction-chains" or absorption of radiation.

Prof. Lindemann suggested to me that there might be enough collisions for unimolecular reactions if the total energy E could be made up by any distribution among a considerable number of degrees of freedom. This note contains some calculations based upon his suggestion. The results are interesting and seem to open the way to a possible new interpretation of unimolecular reactions. The chance of energy in one degree of freedom between E and $E + dE$ is

$$\frac{1}{(\pi RT)^{\frac{1}{2}}} E^{-\frac{1}{2}} . e^{-E/RT} . dE.$$

The chance of a total energy between E and $E + dE$ in n degrees of freedom is

$$\frac{1}{(\pi RT)^{\frac{1}{2}n}} \int_0^E \int_0^E \ldots Q_1^{-\frac{1}{2}} e^{-Q_1/RT} dQ_1 \times Q_2^{-\frac{1}{2}} . e^{-Q_2/RT} . dQ_2 \ldots$$

$$\ldots \{E - (Q_1 + Q_2 \ldots)\}^{-\frac{1}{2}} . e^{-\{E - (Q_1 + Q_2 \ldots)\}/RT} dE,$$

$$= \frac{e^{-E/RT} . E^{(\frac{1}{2}n-1)} . dE}{\Gamma(\frac{1}{2}n) . (RT)^{\frac{1}{2}n}},$$

where $Q_1 + Q_2 + \ldots = E$.

[1] Lindemann, *Trans. Faraday Soc.* **17**, 599 (1922).

(This refers to total kinetic energy; the vibrational degrees of freedom have an equal amount of potential energy. This, however, does not vary independently of the kinetic energy: E can, therefore, be regarded as the "effective" total energy.)

The chance of possessing energy greater than E is

$$\frac{1}{\Gamma(\tfrac{1}{2}n)\,(RT)^{\tfrac{1}{2}n}} \int_{E}^{\infty} e^{-E/RT} . E^{(\tfrac{1}{2}n-1)} . dE.$$

When n is even this integral becomes

$$e^{-E/RT}\left[\frac{1}{\left|\tfrac{1}{2}n-1\right.}\left(\frac{E}{RT}\right)^{\tfrac{1}{2}n-1} + \frac{1}{\left|\tfrac{1}{2}n-2\right.}\left(\frac{E}{RT}\right)^{\tfrac{1}{2}n-2} + \ldots + 1\right].$$

Since E/RT is quite large, we may, for approximate calculations, take the first term only of the expansion, and we have

$$\frac{e^{-E/RT} . (E/RT)^{\tfrac{1}{2}n-1}}{\left|\tfrac{1}{2}n-1\right.}$$

instead of the usual expression $e^{-E/RT}$.

The velocity constant of the reaction will be proportional to this. By taking logarithms and differentiating we obtain

$$d\log k/dT = \frac{E - (\tfrac{1}{2}n-1)\,RT}{RT^2}.$$

Thus to find the real heat of activation the value obtained from the Arrhenius equation must be increased by $(\tfrac{1}{2}n-1)\,RT$. This makes matters less favourable from the point of view of the collision theory.

On the other hand, the factor $\dfrac{(E/RT)^{\tfrac{1}{2}n-1}}{\left|\tfrac{1}{2}n-1\right.}$ may be very large if n is sufficiently great.

[The remainder of this paper is concerned with a quantitative application of the treatment to experimental data on the decomposition of propionaldehyde; this is omitted since it is now known that the decomposition occurs by a complex mechanism. A number of more recent tests of Hinshelwood's formula have shown that it satisfactorily interprets the low-pressure data, but usually the value of n that gives the best fit corresponds to only about half of the normal modes in the molecule.]

THEORIES OF UNIMOLECULAR REACTIONS AT LOW PRESSURES†

O. K. RICE and H. C. RAMSPERGER

[Hinshelwood's paper (Paper 7) successfully dealt with the problem of the pressure range over which the first-order coefficient falls, but it did not provide a completely quantitative interpretation of the form of the variation of the coefficient with the pressure. In the following paper this interpretation is provided by a theory which takes account of the fact that the rate of reaction of the energized molecules is a function of the amount of energy that they contain.

Shortly after this paper appeared L. S. Kassel published a very similar treatment (*J. Phys. Chem.* **32**, 225 (1928)), and he later developed it further in his book *Kinetics of Homogeneous Gas Reactions* (Reinhold, New York, 1932, Chapter 5).

Oscar Knefler Rice (b. 1903) was, at the time this paper was written, at the University of California, and later was at Harvard; he is now Kenan Professor of Chemistry at the University of North Carolina. He has made important contributions to both theoretical and experimental aspects of chemical kinetics.

Herman Carl Ramsperger was born in 1896 and received his Ph.D. degree at the University of California in 1925. Apart from this work with Rice he made a number of experimental and theoretical contributions relating to unimolecular gas reactions. At the time of his death in 1932 he was Assistant Professor at the California Institute of Technology.]

CERTAIN recent experiments on the decomposition of propionic aldehyde[1] show an actual falling off at low pressures in the rate of a reaction which is unimolecular at high pressures. It would

† Most of an article which appeared in the *Journal of the American Chemical Society*, **49**, 1617 (1927).

[1] Hinshelwood and Thompson, *Proc. Roy. Soc.* A, **113**, 221 (1926).

seem that some kind of collision hypothesis must be used to account for activation in such a case. The old form of the collision hypothesis as developed by Langmuir,[2] Christiansen and Kramers,[3] and Tolman[4] appears entirely inadequate even in the case just cited.[5] It seems possible that the explanation may be found in the suggestion, originally made by Lewis and Smith[6] and somewhat extended since by Christiansen[7] and by Hinshelwood and Lindemann,[5] that the internal degrees of freedom may be of some help in causing activation.[8]

This theory may be put in several different forms, not all of which have been considered in this connection. These will be investigated, and equations for the reaction rate at low pressures will be developed, using the methods and results of classical statistical mechanics.

General Considerations

There seems to be some ambiguity over the definition of heat of activation. We shall define it as the minimum energy that a molecule must possess in order that it may decompose. Only internal energy is included and not translational, since we assume with Lewis and Smith[9] that a fast-moving molecule is no more likely to decompose than a slowly-moving one. The heat of activation is *not* to be taken as the excess over an average energy, nor as an energy term derived from the temperature coefficient of the

[2] Langmuir, *J. Amer. Chem. Soc.* **42**, 2201 (1920).

[3] Christiansen and Kramers, *Z. physik. Chem.* **104**, 451 (1923).

[4] Tolman, *J. Amer. Chem. Soc.* **47**, 1524 (1925).

[5] Hinshelwood, *Proc. Roy. Soc.* A, **113**, 230 (1926).

[6] Lewis and Smith, *J. Amer. Chem. Soc.* **47**, 1514 (1925).

[7] Christiansen, *Proc. Cambridge Phil. Soc.* **23**, 438 (1926).

[8] Since this paper was submitted, articles by Fowler and Rideal (*Proc. Roy. Soc.* A, **113**, 570 (1927)) and Thomson (*Phil. Mag.* [7] **3**, 241 (1927)) have been received here, in which these authors have considered the internal degrees of freedom. They were attempting to explain quite a different phenomenon, and many of their assumptions run counter to ours. In so far as their treatments coincide with ours, they are similar also to the earlier work of Hinshelwood and Christiansen.

[9] Lewis and Smith, *J. Amer. Chem. Soc.* **47**, 1512 (1925).

reaction by use of the simple Arrhenius equation. In general, it is to be supposed that the energy of activation may be distributed in any manner whatsoever among the various degrees of freedom, exclusive of those concerned with the motion of translation of the molecule as a whole.

With this understanding we may now turn our attention to various possible theories.

Theory I

The simplest possible hypothesis states that an activated molecule has a certain chance of reacting, independent of the distribution of energy among the degrees of freedom and independent of the amount it has above that necessary for activation. This is the theory which has been considered by Hinshelwood.[5]

Theory II

It might be necessary for reaction that a *particular* degree of freedom should have a minimum energy, say ϵ_0, but that it is able to obtain this energy, even between collisions, from other parts of the molecule.[10] We shall assume that if the proper degree of freedom gets this energy, the molecule instantly decomposes. Now if the molecule *as a whole* has an energy of ϵ_0 or greater, enough of this energy *may* get to the proper degree of freedom to cause reaction to occur. A molecule is *activated*, then, if its total energy, exclusive of the translational, is at least ϵ_0. However, the chance that it will *react* depends on the amount of energy it has in excess of ϵ_0, for this determines the probability that enough energy will reach the right place.

Other Theories

There may be various modifications of Theories I and II. (a) It may be that if *any one* of a number of unconnected degrees of freedom gets a certain (unshared) energy, the reaction occurs.

[10] Thomson (Ref. 8) also makes use of this idea.

(b) Reaction may take place if the energy becomes localized, not in a single degree of freedom, but among a few closely connected degrees of freedom, it being supposed that it makes no difference how it is distributed among this limited number of degrees of freedom. (c) It may be necessary for each of several degrees of freedom simultaneously to attain a minimum energy.

The above are typical, but other cases might readily be imagined. We shall consider chiefly Theories I and II.

It is to be remarked that any of the above will, of course, give a unimolecular reaction at high pressures where there is an equilibrium number of activated molecules, for the rate then depends only on the probability that something will occur within a single activated molecule.

Before we can proceed to the main part of this paper it is necessary to make a general statement concerning the method we propose to use in calculating the rate of activation. As already stated, we shall assume that interchanges of energy between molecules occur only at collisions, and we shall consider only the case in which this is the sole means of activation. Hinshelwood[11] shows that the rate of activation may be calculated from the rate of deactivation. This idea we intend to use, but we desire to go a little further than he does. If the pressure is high only a small fraction of the activated molecules react, most of them becoming deactivated by collisions with other molecules. The rate of activation is then obviously practically equal to the rate of deactivation. The latter may be calculated from the number of collisions of activated molecules, assuming deactivation at every collision. Now, when the pressure is low, it is no longer true that the rate of activation is equal to the rate of deactivation, because an appreciable number of the activated molecules react instead of becoming deactivated; however, this does not affect the rate of activation, and we can calculate the rate of deactivation which would exist if there were an equilibrium number of activated molecules present, and set it equal to the rate of activation. If the assumption of deactivation

[11] Hinshelwood, *Kinetics of Chemical Change in Gaseous Systems*, Oxford, 1926, p. 127.

at every collision is not absolutely true, the last sentence does not hold, for then, by the principle of entire equilibrium, part of the source of activated molecules will be other activated molecules, whose numbers are depleted at low pressures. Now, it may be that although most of the activated molecules lose energy on collision, a small number actually gain it. One might even suppose that the entire supply of the more highly activated molecules (their number being exceedingly small) was maintained in this way from other less highly but nevertheless activated molecules. This would make no difference in the case of Theory I because of the smallness of number of the high-energy molecules. In Theory II, however, the high-energy molecules have much more tendency to react and this may more than counterbalance the other factor. It will, therefore, be necessary to make an *a posteriori* calculation to show that our assumption is justified, or to indicate how much it may lack of being true.

We may now proceed, confining ourselves throughout to the situation existing at the beginning of a reaction, where we may consider that we have a pure gas, the reaction products not yet beginning to have an effect.

Reaction Rate at Low Pressures According to Theory I

Let N be the number of molecules per unit volume, and W the fraction of activated molecules as calculated from the distribution law, assuming equilibrium. NW, then, is the actual number per unit volume of activated molecules at high pressures, but not at low. The number of collisions per unit volume per unit time made by activated molecules at high pressures is aN^2W, where a is a constant known from the kinetic theory of gases. This is also the rate of deactivation per unit volume at high pressures and the rate of activation at any pressure. The rate of deactivation for low pressures is found by considering the number per unit volume, Z, of actually existing activated molecules, and is given by the expression aNZ. The number of molecules decomposing per unit volume per unit time is bZ, where b is a constant. A steady state will exist in which the rate of activation is almost exactly balanced by the combined rates of deactivation and reaction.[12] We then have $aN^2W = aNZ + bZ$. Solving for Z we find

$$Z = aN^2W/(aN + b). \tag{1}$$

Now let K be the fraction of molecules decomposed per second; $K = bZ/N$. Let K_∞ be the fraction decomposed per second at high pressures, or the unimolecular rate constant. $K_\infty = bNW/N = bW$. We have, therefore, substituting for Z, from equation 1, and then for b, from the last equation, in the expression for K, and making algebraic reductions,[13]

$$K = \frac{K_\infty N}{N + K_\infty/(aW)} = \frac{K_\infty p}{p + K_\infty kT/(aW)} \tag{2}$$

where p is the pressure, k the gas constant per molecule and T the absolute temperature.

From the kinetic theory of gases it is known that[14]

$$a = 4s^2\sqrt{\pi kT/m} \tag{3}$$

where s is the molecular diameter and m the mass of a molecule.

From classical statistical mechanics[5]

$$W = \frac{1}{\Gamma\left(\frac{n}{2}\right)} \left(\frac{\epsilon_0}{kT}\right)^{\frac{n-2}{2}} e^{-\frac{\epsilon_0}{kT}} \tag{4}$$

where ϵ_0 is the energy of activation and n is the number of internal degrees of freedom. This is based on the assumption, the equivalent of which is commonly made in the deduction of the law of equipartition of energy,[15] that the energy is expressible as a sum of squares of coordinates and momenta with constant coefficients, or that Hooke's law of force holds within the molecule. This enables us to use total energy in expressions which were originally

[12] Ref. 11, p. 115.
[13] Christiansen (*Trans. Faraday Soc.* **21**, 518 (1926)) also gets an equation of the same general form.
[14] Jeans, *Dynamical Theory of Gases*, Cambridge, 1916, p. 267.
[15] Ref. 11, 1916, ed., pp. 80, 87.

derived for the kinetic energy. In so doing we count one degree of freedom for each coordinate which enters the energy expression, and one for each momentum.

So we obtain finally, by substitution of equations 3 and 4 in 2,

$$K = \frac{K_\infty}{1 + \beta_1/p} \tag{5}$$

where

$$\beta_1 = \frac{K_\infty \sqrt{m} (kT)^{\frac{n-1}{2}} e^{\frac{\epsilon_0}{kT}} \Gamma\left(\frac{n}{2}\right)}{4\sqrt{\pi} s^2 \epsilon_0^{\frac{n-2}{2}}}.$$

It is to be noted that the temperature coefficient of the rate constant is related to the heat of activation by the equations[5]

$$\frac{d \log K_\infty}{dT} = \frac{d \log W}{dT} = \frac{\epsilon_0 - \frac{n-2}{2} kT}{kT^2} = \frac{U}{kT^2} \tag{6}$$

where U would be the energy of activation per molecule, as originally defined by Arrhenius and used by many writers in this field.

Reaction Rate at Low Pressures According to Theory II

This case differs from the former chiefly in that we must consider the activated molecules with different amounts of energy separately. We will let $W_\epsilon d\epsilon$ be the fraction of molecules having total internal energy between ϵ and $\epsilon + d\epsilon$ as calculated from the normal distribution law for the case of equilibrium, so that

$$W = \int_{\epsilon_0}^{\infty} W_\epsilon d\epsilon, \; W \text{ having the same meaning as before. We will}$$

assume that the number of molecules per unit volume per unit time entering the range between ϵ and $\epsilon + d\epsilon$ is given by $aN^2 W_\epsilon d\epsilon$, the number which would leave that range if equilibrium existed (that is, the number of collisions). The justification of this we will defer, as explained above.

We will let $Z_\epsilon d\epsilon$ be the *actual* number of molecules per unit volume in the range between ϵ and $\epsilon + d\epsilon$, so that $Z = \int_{\epsilon_0}^{\infty} Z_\epsilon d\epsilon$. The number per unit volume per unit time leaving the range through deactivation will then be $aNZ_\epsilon d\epsilon$, while the number lost through reaction will be $b_\epsilon Z_\epsilon d\epsilon$; b_ϵ will differ for different energies in a way to be determined.

We then have $aN^2 W_\epsilon d\epsilon = aNZ_\epsilon d\epsilon + b_\epsilon Z_\epsilon d\epsilon$, so

$$Z_\epsilon = \frac{aN^2 W_\epsilon}{aN + b_\epsilon}. \tag{7}$$

Now let $K_\epsilon d\epsilon$ be the total number of molecules in the range ϵ to $\epsilon + d\epsilon$ which decompose per second, divided by the total number of molecules. Then, $K_\epsilon d\epsilon = b_\epsilon Z_\epsilon d\epsilon / N = aNW_\epsilon b_\epsilon d\epsilon / (aN + b_\epsilon)$ by equation 7, and

$$K = \int_{\epsilon_0}^{\infty} K_\epsilon d\epsilon = \int_{\epsilon_0}^{\infty} \frac{W_\epsilon b_\epsilon d\epsilon}{1 + b_\epsilon kT/(ap)} \tag{8}$$

where p/kT has been substituted for N. It will be noted that for high pressures b_ϵ is negligible compared to aN and we have

$$K_\infty = \int_{\epsilon_0}^{\infty} W_\epsilon b_\epsilon d\epsilon. \tag{9}$$

By the general principles of statistical mechanics[16] under the same assumptions as before

$$W_\epsilon d\epsilon = \frac{1}{\Gamma\left(\dfrac{n}{2}\right)} \left(\frac{\epsilon}{kT}\right)^{\frac{n-2}{2}} e^{-\frac{\epsilon}{kT}} \frac{d\epsilon}{kT} \tag{10}$$

We now have to evaluate b_ϵ.

Let us consider a number of molecules, all of which have energy between ϵ and $\epsilon + d\epsilon$, ϵ being greater than ϵ_0. These we shall designate as molecules of Class A. There will then be a certain chance that a molecule of Class A will have an energy larger than

[16] Gibbs, *Elementary Principles in Statistical Mechanics*, Scribner's, New York, 1902, p. 93.

ϵ_0 in a particular degree of freedom. We will call such a molecule a molecule of Class B. If a state of equilibrium existed, these molecules would be shifting over at a certain rate (between collisions) to molecules which, while still of Class A, would not be of Class B. Let us suppose that this rate is proportional to the number of Class B, and is independent (1) of the excess of energy in the particular degree of freedom[17] and (2) of ϵ.[18] Now the rate at which molecules of Class B are formed under equilibrium conditions is equal to the rate at which they revert to molecules of Class A and not of Class B, and hence is proportional to the number of Class B. If all molecules of Class B decompose, this does not affect the rate at which they are formed. Hence the rate of decomposition is proportional to the number of Class B which would normally exist in a state of equilibrium.

We have, then, to determine the fraction of all molecules which have total energy between ϵ and $\epsilon + d\epsilon$ and which would also have, if the distribution law held, one degree of freedom with energy greater than ϵ_0.

The chance that one particular degree of freedom has an energy between ϵ_1 and $\epsilon_1 + d\epsilon_1$ is

$$\frac{1}{\sqrt{\pi}} \left(\frac{\epsilon_1}{kT} \right)^{-\frac{1}{2}} e^{-\frac{\epsilon_1}{kT}} \frac{d\epsilon_1}{kT}. \tag{11}$$

The chance that all the rest of the degrees of freedom have among them an energy between $\epsilon - \epsilon_1 - d\epsilon_1$ and $\epsilon - \epsilon_1 + d\epsilon$ is

$$\frac{1}{\Gamma\left(\frac{n-1}{2} \right)} \left(\frac{\epsilon - \epsilon_1}{kT} \right)^{\frac{n-3}{2}} e^{-\frac{\epsilon - \epsilon_1}{kT}} \frac{d\epsilon + d\epsilon_1}{kT}. \tag{12}$$

---

[17] This certainly cannot be a bad assumption to make for purposes of approximation, for the distribution law shows that the number of molecules having any appreciable excess over ϵ_0 in the particular degree of freedom is very small compared with the number having a slight excess, and hence those with a large excess do not greatly affect the calculation.

[18] The shape of the curves K_ϵ against $\epsilon - \epsilon_0$ later to be obtained (Fig. 8.1) would indicate that this should be satisfactory for purposes of approximation, unless the effect of ϵ is quite marked.

The product of equations 11 and 12 is evidently greater than the probability that a molecule should have energy between ϵ and $\epsilon + d\epsilon$ and at the same time should have an energy between ϵ_1 and $\epsilon_1 + d\epsilon_1$ in one (specified) degree of freedom.[19] Similarly, if we had written $d\epsilon - d\epsilon_1$, instead of $d\epsilon + d\epsilon_1$ (assuming $d\epsilon_1 < d\epsilon$), then their product would have been smaller. If $d\epsilon_1$ is small compared with $d\epsilon$, then it does not matter which way equation 12 is written, and we may write it simply

$$\frac{1}{\Gamma\left(\dfrac{n-1}{2}\right)} \left(\frac{\epsilon - \epsilon_1}{kT}\right)^{\frac{n-3}{2}} e^{-\frac{\epsilon - \epsilon_1}{kT}} \frac{d\epsilon}{kT}. \tag{13}$$

It is now evident that the product of equations 11 and 13 may be integrated with respect to ϵ_1, holding ϵ and $d\epsilon$ constant. This removes the limitation that $d\epsilon_1$ must be small compared to $d\epsilon$.[20]

Multiplying equations 11 and 13 and integrating with respect to ϵ_1 from ϵ_0 to ϵ will give us $aW_\epsilon b_\epsilon$, where a is a constant, independent of ϵ, which will be determined later.

$$aW_\epsilon b_\epsilon = \frac{1}{\sqrt{\pi}\,\Gamma\left(\dfrac{n-1}{2}\right)} \frac{e^{-\frac{\epsilon}{kT}}}{(kT)^{\frac{n}{2}}} \int_{\epsilon_0}^{\epsilon} \epsilon_1^{-\frac{1}{2}} (\epsilon - \epsilon_1)^{\frac{n-3}{2}} d\epsilon_1. \tag{14}$$

For any definite value of n the integral in this expression could be evaluated by successive integration by parts. We will, however, be content with an approximation. We see that $(\epsilon - \epsilon_1)^{(n-3/2)}$ varies much more rapidly, assuming n to be fairly large, than

[19] This is seen in the following manner. Suppose the one degree of freedom to have an energy exactly ϵ_1. Then if all the other degrees of freedom together can have an energy between $\epsilon - \epsilon_1 - d\epsilon_1$ and $\epsilon - \epsilon_1 + d\epsilon$, the total energy of the molecule can lie anywhere between $\epsilon - d\epsilon_1$ and $\epsilon + d\epsilon$. This evidently includes every molecule in which the one degree of freedom has an energy of exactly ϵ_1 and in which the total energy is between ϵ and $\epsilon + d\epsilon$, and some others besides. The same can be seen to be true for any value we choose to take between ϵ_1 and $\epsilon_1 + d\epsilon_1$.

[20] For treatment of a similar problem see Lorentz, *Théories Statistiques en Thermodynamique*, Teubner, Leipzig and Berlin, 1916, p. 12.

$\epsilon_1{}^{-1/2}$, and that the important part of the integrand is the part where ϵ_1 is small. We therefore write as an approximation

$$\int_{\epsilon_0}^{\epsilon} \epsilon_1^{-\frac{1}{2}}(\epsilon-\epsilon_1)^{\frac{n-3}{2}} d\epsilon_1 = \epsilon_0^{-\frac{1}{2}}\int_{\epsilon_0}^{\epsilon}(\epsilon-\epsilon_1)^{\frac{n-3}{2}} d\epsilon_1 = \frac{2\epsilon_0^{-\frac{1}{2}}(\epsilon-\epsilon_0)^{\frac{n-1}{2}}}{n-1}.$$
(15)

To evaluate α we substitute equation 15 in 14 and integrate with respect to ϵ from ϵ_0 to ∞. This gives us

$$\alpha\int_{\epsilon_0}^{\infty} W_\epsilon b_\epsilon d\epsilon = \frac{2\epsilon_0^{-\frac{1}{2}} e^{-\frac{\epsilon_0}{kT}} (kT)^{\frac{1}{2}}\Gamma\left(\frac{n+1}{2}\right)}{(n-1)\sqrt{\pi}\,\Gamma\left(\frac{n-1}{2}\right)}.$$
(16)

Evaluating α from equations 16 and 9 and inserting the value in equation 14 we have

$$W_\epsilon b_\epsilon = \frac{K_\infty}{\Gamma\left(\frac{n+1}{2}\right)} e^{-\frac{\epsilon-\epsilon_0}{kT}} \frac{1}{kT}\left(\frac{\epsilon-\epsilon_0}{kT}\right)^{\frac{n-1}{2}}.$$
(17)

From equations 17 and 10

$$b_\epsilon = \frac{K_\infty\,\Gamma\left(\frac{n}{2}\right)}{\Gamma\left(\frac{n+1}{2}\right)} \frac{e^{\frac{\epsilon_0}{kT}}}{(kT)^{\frac{1'}{2}}} \frac{(\epsilon-\epsilon_0)^{\frac{n-1}{2}}}{\epsilon^{\frac{n-2}{2}}}.$$
(18)

Equation 8 now becomes

$$K = \frac{K_\infty}{kT\,\Gamma\left(\frac{n+1}{2}\right)} \int_0^{\infty} \frac{e^{-\frac{\epsilon-\epsilon_0}{kT}}\left(\frac{\epsilon-\epsilon_0}{kT}\right)^{\frac{n-1}{2}}}{1+\frac{\beta_2}{p}\frac{(\epsilon-\epsilon_0)^{\frac{n-1}{2}}}{\epsilon^{\frac{n-2}{2}}}} d(\epsilon-\epsilon_0)$$
(19)

where

$$\beta_2 = K_\infty \frac{\Gamma\left(\frac{n}{2}\right)}{\Gamma\left(\frac{n+1}{2}\right)} \sqrt{\frac{m}{\pi}} \frac{1}{4s^2} e^{\frac{\epsilon_0}{kT}}.$$

This has been integrated by graphical means, and a series of curves for this purpose of the integrand, K_ϵ, against $\epsilon - \epsilon_0$ is shown in Fig. 8.1. The outer curve is the curve for infinite pressure, and the number near each curve gives the value of β'/p, where $\beta' = 3 \cdot 16 \times 10^{-3}\beta_2$, ϵ and ϵ_0 being in calories per mole. The

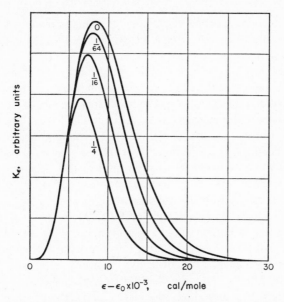

FIG. 8.1.—Curve $\beta'/p = 0$, relative area $(K/K\infty) = 1 \cdot 00$; $\beta'/p = {}^1/_{64}$, $(K/K\infty) = 0 \cdot 84$; $\beta'/p = {}^1/_{16}$, $(K/K\infty) = 0 \cdot 66$; $\beta'/p = {}^1/_4$, $(K/K\infty) = 0 \cdot 44$.

temperature used was 576°, n was taken as 11, and ϵ_0 as 54,000 calories per mole. These figures correspond to some of Hinshelwood and Thompson's experiments on propionic aldehyde. The ordinates represent the relative number of molecules having a definite energy and pressure which react in a given time. Due to the character of the function the value of the integrand decreases rapidly with high values of $\epsilon - \epsilon_0$, and although the curve theoretically extends to infinity, actually we can use a finite portion of it.

In evaluating the heat of activation from the temperature coefficient of the reaction rate in the case of Theory II, we note that here

$$\frac{d \log K_\infty}{dT} = \frac{\epsilon_0 + \frac{1}{2}kT}{kT^2} = \frac{U}{kT^2} \tag{20}$$

as shown by Rodebush[21] since the rate at high pressures depends on the probability that a *single* degree of freedom has an energy greater than ϵ_0.

We can now use the set of curves in Fig. 8.1 to justify the assumption we made that the actual rate of formation of molecules in a given energy range was to be calculated from the rate of loss which would exist under equilibrium conditions. The lower the pressure the less justified we would consider this assumption to be. Consider, therefore, the lowest curve of the set shown. A trifle over 11% of the area under it is to the right of the 11,000 mark. At $\epsilon - \epsilon_0$ equal to 5000 the concentration is about 8% of the equilibrium value. We now ask the question: How many of the molecules having an energy per mole of 65,000 cal ($\epsilon - \epsilon_0 = 11,000$) were formed from molecules having an energy of more than 59,000 cal per mole, and hence not present in approximately equilibrium numbers? To answer this let us use the principle of entire equilibrium and find what is the chance that a molecule of 65,000 cal per mole energy should not lose more than 6000 cal per mole at a collision with an average molecule (or that the other molecule should leave with more than 59,000 cal per mole). We can do this if we assume that statistical equilibrium is established between the various degrees of freedom of both molecules at a collision. If statistical equilibrium is not established at a collision, the error would be about the same for low- and high-energy molecules, and the chief result would be that only a fraction of the collisions would be effective in deactivation, which would merely mean that β_2/p would have to be multiplied by a constant factor. While this last point cannot be said to be strictly established, it is obvious that the

[21] Rodebush, *J. Amer. Chem. Soc.* **45**, 610 (1923).

calculation we propose to make will be a valuable indication of the validity of our assumption.

If two molecules, each having n degrees of freedom, collide, there are $2n$ degrees of freedom involved. The chance that one of the molecules with n degrees of freedom should have an energy between ϵ_2 and $\epsilon_2 + d\epsilon_2$ is

$$\frac{1}{\Gamma\left(\dfrac{n}{2}\right)} \left(\frac{\epsilon_2}{kT}\right)^{\frac{n-2}{2}} e^{-\frac{\epsilon_2}{kT}} \frac{d\epsilon_2}{kT}. \tag{21}$$

The chance that the other molecule should have energy between $\epsilon' - \epsilon_2$ and $\epsilon' - \epsilon_2 + d\epsilon'$ is

$$\frac{1}{\Gamma\left(\dfrac{n}{2}\right)} \left(\frac{\epsilon' - \epsilon_2}{kT}\right)^{\frac{n-2}{2}} e^{-\frac{\epsilon' - \epsilon_2}{kT}} \frac{d\epsilon'}{kT}. \tag{22}$$

By reasoning similar to that following equations 12 and 13, the product of 21 and 22 is the probability that the total energy of the two molecules is between ϵ' and $\epsilon' + d\epsilon'$, while one of them has energy between ϵ_2 and $\epsilon_2 + d\epsilon_2$. But the chance that the two together have energy between ϵ' and $\epsilon' + d\epsilon'$ is

$$\frac{1}{\Gamma(n)} \left(\frac{\epsilon'}{kT}\right)^{n-1} e^{-\frac{\epsilon'}{kT}} \frac{d\epsilon'}{kT}. \tag{23}$$

The product of equations 21 and 22 divided by equation 23 is the chance that if two molecules have an energy of ϵ' together, a certain one of them has an energy between ϵ_2 and $\epsilon_2 + d\epsilon_2$. Now we start with two molecules, one of which has an energy of 65,000 and the other the average energy or 9300 cal per mole, the sum of which is ϵ'. We wish to find the chance that either of them should come out of the collision with an energy greater than 59,000 cal per mole (ϵ''). We must double the expression $[(21) \times (22)/(23)]$ to allow for either molecule having the energy, and integrate.

$$\frac{2\Gamma(n)}{\epsilon'^{\frac{n}{2}}\left[\Gamma\left(\dfrac{n}{2}\right)\right]^2} \int_{\epsilon''}^{\epsilon'} \left(\frac{\epsilon_2}{\epsilon'}\right)^{\frac{n-2}{2}} (\epsilon' - \epsilon_2)^{\frac{n-2}{2}} \, d\epsilon_2. \tag{24}$$

For the sake of simplicity, and since it can only cause the probability to come out greater than it should be, we will take ϵ_2/ϵ' as 1, and get

$$\frac{4\Gamma(n)}{n\left[\Gamma\left(\dfrac{n}{2}\right)\right]^2} \left(\frac{\epsilon'-\epsilon''}{\epsilon'}\right)^{\frac{n}{2}} = 0\cdot081. \tag{25}$$

This seems small enough to justify our assumption as an approximation, especially when it is considered that the molecule we fixed attention on had considerably higher energy than the average reacting molecule.

Summary

A collision theory of homogeneous gas-phase, unimolecular reactions at low pressures has been developed. Several possible forms of the theory are suggested, and two of them carefully considered. These theories take account of the distribution of energy among the large number of degrees of freedom of the rather complex molecules that are known to decompose in unimolecular manner. They determine the way in which the reaction rate would be influenced by the initial pressure and can, therefore, be tested experimentally. All molecules which have a certain minimum total energy which can be calculated from the temperature coefficient are said to be activated. The rate of activation is calculated from the distribution of energy among molecules having n degrees of freedom, and by assuming that the rate of activation is equal to what the rate of deactivation would be under equilibrium conditions, and that each collision of an activated molecule results in deactivation.

When the pressure is high, most activated molecules are deactivated and but a small fraction react. At lower pressures an appreciable fraction of them react, and at very low pressures when most of the activated molecules react the reaction rate becomes very much smaller and bimolecular. By Theory I, the chance of reaction is independent of the excess of energy beyond that required for activation, and this leads to an equation of the form $K = K_\infty p/(p+\beta_1)$, where K is the rate constant, K_∞ the rate constant at high pressures, p the pressure and β_1 a constant involving the number of degrees of freedom, the molecular diameter, the heat of activation, K_∞, etc.

Theory II requires not only that a molecule have a minimum total energy, but that such an activated molecule react only if it contains a certain minimum of energy in a particular degree of freedom. It is assumed that the internal energy may be repeatedly redistributed within the molecules between collisions. If a molecule has a large excess of energy beyond that required for activation, there will be a greater probability that this energy will get into a particular degree of freedom to cause reaction. An equation is derived which gives this probability. By fixing the number of degrees of freedom, etc., it is possible to determine the relative reaction rate at various pressures by means of a graphical integration.

The data on the decomposition of propionic aldehyde at various pressures are found to fit the theoretical curves of Theory I and Theory II about equally well. More accurate data extending over a wider pressure range are necessary to decide between the two theories.

[Two pages of the paper, dealing with applications of the treatment to the propionaldehyde decomposition, have been omitted; this reaction is now known to be complex. Subsequent work has shown that the Rice–Ramsperger–Kassel treatment gives a fairly satisfactory interpretation of the fall-off in rate coefficient at low pressures, for a number of reactions. However, the value of n, the number of degrees of freedom, required to give agreement with experiment, is usually much less than the total number of vibrational degrees of freedom in the molecule. A quantum version of the treatment, which takes into account zero-point energies and does not assume all vibrational frequencies to be the same, has been proposed by Marcus and Rice (R. A. Marcus and O. K. Rice, *J. Phys. Colloid Chem.* **55**, 894 (1951); G. M. Weider and R. A. Marcus, *J. Chem. Phys.* **37**, 1835, (1962), and has proved more satisfactory than the original Rice–Ramsperger–Kassel versions.]

PART IV

Chain Reactions

ON THE REACTION BETWEEN HYDROGEN AND BROMINE†

J. A. Christiansen

[The reactions of hydrogen with the halogens were among the first gas-phase reactions to be thoroughly studied kinetically. An empirical expression for the rate of formation of hydrogen bromide from hydrogen and bromine had been proposed by M. Bodenstein and S. C. Lind (*Z. physik. Chem.* **57**, 168 (1907)). The present paper by Christiansen gives a clear-cut explanation of this expression in terms of the mechanism of the reaction. In this paper the ideas of chain initiation, termination and propagation are related specifically to the hydrogen–bromine reaction, and the empirical expression for the rate is shown to follow from the mechanism proposed.

A word of explanation may clarify the development of equation (7). Implicit in the argument is the assumption of a steady-state concentration of hydrogen atoms:

$$\frac{d[\mathrm{H}]}{dt} = k_2[\mathrm{Br}][\mathrm{H_2}] - k_3[\mathrm{H}][\mathrm{Br_2}] - k_4[\mathrm{H}][\mathrm{HBr}]$$

$$[\mathrm{H}] = \frac{k_2[\mathrm{Br}][\mathrm{H_2}]}{k_3[\mathrm{Br_2}] + k_4[\mathrm{HBr}]} = \frac{P'[\mathrm{Br}]}{P_3 + P_4}.$$

The rate of production of HBr may be expressed as

$$\frac{d[\mathrm{HBr}]}{dt} = k_2[\mathrm{Br}][\mathrm{H_2}] + k_3[\mathrm{H}][\mathrm{Br_2}] - k_4[\mathrm{H}][\mathrm{HBr}]$$

$$= P'[\mathrm{Br}] + \frac{P'[\mathrm{Br}]}{P_3 + P_4}[P_3 - P_4]$$

this equation being equation (7) of the paper.

The following excerpt forms Part I of Christiansen's paper. Parts II and III are concerned with a calculation of the energy of the reaction $\mathrm{Br} + \mathrm{HBr} = \mathrm{Br_2} + \mathrm{H}$ and the consequence of this in relation to the heat of dissociation of hydrogen; since this approach is not as relevant to present-day kinetics these parts are not included.

---

† Part I of a paper which appeared, in English, in *Det. Kgl. Danske Viben-skabernes Selskab. Mathematisk-fysiske Meddelelser*, **1**, 14 (1919).

J. A. Christiansen was born in Denmark in 1888. This paper was written before he received his doctorate degree and while he was assistant in the Chemical Laboratory of the University of Copenhagen. He eventually became Professor of Chemistry and Head of the Department of Physical Chemistry at this University and remained there until his retirement in 1959. He has received many honours in recognition of his important contributions in the field of reaction kinetics.]

M. BODENSTEIN and S. C. Lind[1] have shown that the (isothermal) reaction

$$H_2 + Br_2 = 2HBr$$

goes on according to the equation

$$\frac{dx}{dt} = k_1 \frac{(a-x)(b-x)^{\frac{1}{2}}}{m + \dfrac{x}{b-x}}.$$

Here they denote by a and b the original concentrations of hydrogen and bromine respectively, and by $2x$ the concentration of the hydrogen bromide formed. Consequently the equation can be rewritten

$$\frac{d\,C_{HBr}}{dt} = 2k_1 \frac{C_{H_2} \cdot C_{Br_2}^{\frac{1}{2}}}{2m + \dfrac{C_{HBr}}{C_{Br_2}}}. \tag{1}$$

This equation is the empirical expression of the facts found by these authors. The denominator tells us that the reaction-velocity decreases with increasing concentration of hydrogen bromide, but not below a certain limit, and further that it is the *ratio* between the concentrations of hydrogen bromide and bromine, not the absolute value of the first, that determines the velocity-decrement.

The authors have further shown that such gases as carbon tetrachloride, air, and steam have no influence on the velocity, while iodine has a still greater action.

We will now show that by the aid of these data we can draw very definite conclusions concerning the mechanism of the reaction considered.

[1] M. Bodenstein and S. C. Lind, *Z. Physik. Ch.* **57**, 168 (1907).

Let us suppose—in accordance with Bodenstein and Lind—that the bromine also by these relatively low temperatures ($225°-301°C$) to a certain degree is dissociated in atoms. This assumption is in concordance with the form of the numerator in equation (1) when we further suppose the primary reaction to be

$$Br + H_2 = HBr + H. \tag{2}$$

But from a number of facts[2] we know that an atom of hydrogen forms a very reactive system, and consequently we may expect it to react in one of the following five ways:

(1) $H + H = H_2$ (2) $H + Br = HBr$ (3) $H + H_2 = H_2 + H$
(4) $H + Br_2 = HBr + H$ (5) $H + HBr = H_2 + Br$

in a time which is comparable with the time of describing a free path. The possibilities (1) and (2) will occur extremely seldom on account of the small concentrations of the atoms, while the occurrence of (3), a possibility which is not altogether to be denied, will make no alterations at all in the state of the whole system. The only two ways left for the hydrogen-atom to react in are consequently

$$H + Br_2 = HBr + Br \tag{3}$$

or $\qquad H + HBr = H_2 + Br. \tag{4}$

As is well known, the equilibrium $2Br \rightleftharpoons Br_2$ is reached almost instantly, and so the atoms of bromine formed by the reactions (3) and (4), or those used up by the reaction (2), cannot alter the concentration of atoms of bromine (at constant C_{Br_2}).

If now we denote by P_3 and P_4 respectively the relative probabilities for the occurrences of the reactions (3) and (4), we can write

$$P_3 + P_4 = 1, \tag{5}$$

$$\frac{P_3}{P_4} = K \frac{C_{Br_2}}{C_{HBr}} \tag{6}$$

where K probably depends on the temperature but not on C_{Br_2} and C_{HBr}.

---

[2] Compare, for instance, I. Langmuir, *J. Amer. Chem. Soc.* **36**, 711 (1914).

E

If we further let P' and P'' respectively denote the probabilities per unit time for an atom of bromine to react with a hydrogen-molecule according to (2), or with a molecule of hydrogen bromide according to (9) (see below), the velocity of the formation of hydrogen bromide must be

$$\frac{d\ C_{\text{HBr}}}{dt} = C_{\text{Br}} \cdot P'(1 + P_3 - P_4) \tag{7}$$

and the velocity of the opposing reaction

$$-\frac{d\ C_{\text{HBr}}}{dt} = C_{\text{Br}} \cdot P''(1 + P_4 - P_3). \tag{8}$$

In the latter equation we have supposed that the primary reaction by the dissociation is

$$\text{Br} + \text{HBr} = \text{Br}_2 + \text{H}. \tag{9}$$

This assumption is a necessary consequence of our theory concerning the mechanism of formation, and of the requirement that the condition of equilibrium shall be of the well-known form.

P' and P'' are determined by

$$P' = k'C_{\text{H}_2} \tag{10}$$

and $\qquad\qquad P'' = k''C_{\text{HBr}}. \tag{11}$

In these equations we denote by k' and k'' the velocity constants defined in the usual way, and so they may be expected to depend on the temperature, but not on the concentrations.

From the equations (7), (8) and (5) we get the total velocity of formation

$$\frac{d\ C_{\text{HBr}}}{dt} = 2C_{\text{Br}}[P'(1 - P_4) - P''\ P_4]$$

and by elimination of P', P'' and P_4

$$\frac{d\ C_{\text{HBr}}}{dt} = 2C_{\text{Br}}\left(K\,k'\ C_{\text{H}_2}\frac{C_{\text{Br}_2}}{C_{\text{HBr}}} - k''\ C_{\text{HBr}}\right) : \left(1 + K\frac{C_{\text{Br}_2}}{C_{\text{HBr}}}\right)$$

$$= 2C_{\text{Br}}\left(K\,k'\ C_{\text{H}_2} - k''\ \frac{C_{\text{HBr}}}{C_{\text{Br}_2}}\right) : \left(\frac{C_{\text{HBr}}}{C_{\text{Br}_2}} + K\right). \tag{12}$$

Now by the experiments considered k'' is vanishing compared with k', and as we furthermore have

$$\frac{C_{Br}^2}{C_{Br_2}} = K_b, \tag{13}$$

equation (12) can be rewritten

$$\frac{d\,C_{HBr}}{dt} = 2\,K\,k'\,K_b^{\frac{1}{2}}\,\frac{C_{H_2}\cdot C_{Br_2}^{\frac{1}{2}}}{K + \dfrac{C_{HBr}'}{C_{Br_2}}}. \tag{14}$$

It is evident that equations (1) and (14) are identical in form, and so we have got an explanation of the empirical equation of Bodenstein and Lind.

The action of iodine vapour cannot be explained in an analogous way. But it seems quite natural to me to suppose, as the said authors[3] have a little hesitatingly done, that added iodine diminishes the concentration of bromine molecules and consequently also of bromine atoms in forming IBr, especially when we remember that the iodine in these experiments was added in considerable excess.

Comparing (1) and (14) we get

$$K = 2m \tag{15}$$

and

$$k' = \tfrac{1}{2}k_1 : \left(mK_b^{\frac{1}{2}}\right). \tag{16}$$

The authors have determined k_1 by a series of temperatures. Whether m depends on the temperature can hardly be decided from the experiments in question. The authors use the value 5.00 for the temperatures investigated.[4] It is determined, as far as can be seen, from the experiments at $301\cdot3°C$.

It is interesting to note that this theory is founded on the kinetic measurements alone, and without use of Warburg's results concerning the photolysis of hydrogen bromide[5] which were unknown to me when I first worked it out, and yet the agreement is

[3] loc. cit., p. 188.
[4] loc. cit., p. 180.
[5] Warburg, Berichte Berliner Akad. 1916 (I), p. 314.

complete. Warburg has shown, not only that hydrogen-atoms formed by the primary reaction

$$HBr = H + Br$$

react with hydrogen bromide, according to equation (4), but it also appears from his experiments that the reaction

$$H + Br_2 = HBr + Br \tag{3}$$

goes on, since he has found that the quantity of hydrogen bromide dissociated per unit time depends on the velocity of the gas-current passing the insolation cell. This phenomenon he ascribes to organic matter used in the building up of the cell, but after the theory here developed it seems doubtless that the reaction (3) is also active in this direction.

If we suppose it to be the only significant cause it is possible in this way to determine K independently of the kinetic measurements. If by n we denote the number of hydrogen-atoms primarily formed, and by M the number of hydrogen-bromide molecules disappeared, we get

$$M = n(1 + P_4 - P_3) = 2nP_4$$

and again

$$M\left(K + \frac{C_{HBr}}{C_{Br_2}}\right) = 2n\frac{C_{HBr}}{C_{Br_2}}. \tag{17}$$

From this equation it is possible to determine K if $\dfrac{C_{HBr}}{C_{Br_2}}$ is small compared with K. For great values of this fraction equation (17) is equivalent with

$$M = 2n, \tag{17a}$$

which Warburg has shown to be in accordance with experiment, when n is determined by the aid of Einstein's law on the photochemical equivalent.

Finally we will try to apply quite analogous considerations to the formation and dissociation of hydrogen iodide, although these reactions according to Bodenstein[6] are regularly bimolecular.

[6] Bodenstein, *Z. Physik. Ch.* **20**, 295 (1899).

We can transform (12) in

$$\frac{d\,C_{HI}}{dt} = \frac{2C_I}{C_{HI} + KC_{I_2}}\,(K\,k'C_{H_2}\cdot C_{I_2} - k''\,C_{HI}^2)$$

and, when we put $C_I = K_I^{\frac{1}{2}}\cdot C_{I_2}^{\frac{1}{2}}$ further in

$$\frac{d\,C_{HI}}{dt} = \frac{2\,K_I^{\frac{1}{2}}}{\dfrac{C_{HI}}{C_{I_2}^{\frac{1}{2}}} + KC_{I_2}^{\frac{1}{2}}}\,(K\,k'\,C_{H_2}\cdot C_{I_2} - k''C_{HI}^2).$$

Now the quantity $\dfrac{C_{HI}}{C_{I_2}^{\frac{1}{2}}} + KC_{I_2}^{\frac{1}{2}}$, has a minimum (when $\dfrac{C_{HI}}{C_{I_2}} = K$),

and consequently it may be difficult to see whether the factor outside the parenthesis is necessary to express the results exactly or not, especially as it seems to be rather difficult to determine the constants with any great accuracy.

It follows from the above equation that, when C_{I_2} is very small, we should have

$$\frac{d\,C_{HI}}{dt} = -2\,K_I^{\frac{1}{2}}\cdot k''\cdot C_{HI}\cdot C_{I_2}^{\frac{1}{2}},$$

i.e. the reaction should be autocatalysed by the iodine formed.

That it is really so, is suggested by a curve from Bodenstein's work on the dissociation of hydriodic acid.[7] Inspection shows that without a point of inflection a curve cannot be drawn through the points found by experiments (dissociation curve) and the zero point of the coordinates.

If now these points were determined very exactly, this would definitely show that the said explanation was the right one. Unfortunately however, it seems that the results are hardly sufficient to draw this conclusion.

The same author has found[8] that the photolysis of hydrogen iodide is simply monomolecular, but as the vapour-pressure of iodine is very small at ordinary temperature, and consequently the

ratio $\dfrac{C_{HI}}{C_{I_2}}$ great, this gives us no other information as to the

[7] Bodenstein, *Z. Physik. Ch.* 13, 111 (1894).
[8] Bodenstein, *Z. Physik. Ch.* 22, 23 (1897).

mechanism of the process (compare equations (17) and (17a)), than this, that the reactions

$$H + I_2 = HI + I \text{ and } H + HI = H_2 + I$$

are both possible as secondary processes, when the primary one is supposed to be

$$HI = H + I.$$

In an earlier paper Warburg[9] has shown the occurrence of the last-named two reactions.

[9] Warburg, *Ber. Berliner Akademie* 300 (1918).

THE OXIDATION OF PHOSPHORUS VAPOUR AT LOW PRESSURES†

N. Semenoff

[Certain reactions, such as the hydrogen–oxygen and phosphorus–oxygen reactions, have the remarkable feature that as the pressure is varied their rates may change sharply from a low to a very high value. The first indication of this appears to have been obtained by Chariton and Walta, and was later confirmed by Semenoff, Hinshelwood and their collaborators. The paper that is reproduced below is the first one in which the explosion limits were explained on the basis of a reaction mechanism.

Nikolaj Nikolajevity Semenoff was born in Russia in 1896, and at the time this paper was written was Director of Chemical Physics of the Physico-Chemical Institute in Leningrad. He later became Head of the Institute for Chemical Physics of the Academy of Sciences in Moscow, and Head of the Department of Chemical Kinetics at the University of Moscow. He has done a great deal of important work particularly in the field of the kinetics of gas-phase reactions, and shared the 1956 Nobel Prize in Chemistry with C. N. Hinshelwood.]

Abstract

Chariton and Walta have established that phosphorus vapour does not react at all with oxygen if the pressure of the latter falls below a certain value p_k. In a brief note Bodenstein has sharply criticized the techniques of the above work and has denied that the existence of a critical pressure has been proved. In the present article it will first be discussed that the existence of a critical pressure can be concluded from the work by Chariton and Walta even if Bodenstein's objections

---

† A portion, translated from the German, of an article which appeared in *Zeitschrift für Physik*, **46**, 109 (1927).

are accepted as valid; from the latter it only follows that the numerical results for the critical pressure are inaccurate owing to the method of measurement. The main content of the present article concerns new experiments on the gas reaction between phosphorus and oxygen in which the earlier experimental technique, objected to by Bodenstein and also found by us to be unsuitable, is replaced by another one. We believe to have proved unambiguously the dependence of the critical oxygen pressure v_k, (a) on the pressure of the phosphorus (p_{P_4}), (b) on the pressure of the added argon (p_{Ar}), (c) on the temperature and (d) on the diameter of the reaction vessel. It has been found that the critical pressure decreases with an increase of the pressure p_{P_4} and p_{Ar}; the same is the case when the diameter of the vessel is increased. The critical pressure is independent of the temperature. The results obtained can be interpreted qualitatively on the basis of the following concepts: (a) A direct reaction between the molecules P_4 and O_2 is impossible. (b) There are reacting some active molecules, which lose their activity when colliding with the walls of the vessel. (c) The reaction has a chain character, but it proceeds very slowly owing to the small number of the initiating centres; however, under certain conditions it can go over into an explosion as has also been calculated by Christiansen and Kramers.

CHARITON and Walta[1] have studied in the laboratory the oxidation of phosphorus at low pressures under the following conditions: the oxygen was supplied through a capillary tube to a previously evacuated bulb containing phosphorus. A McLeod gauge was connected to the bulb, and between the bulb and the manometer was situated a liquid-air trap; this prevented penetration of the phosphorus into the McLeod gauge and of the mercury into the bulb.

In this way the following phenomena were observed: the flash occurred in the bulb not immediately after the admission of the oxygen through the capillary tube but after a certain interval of time which, depending on the rate of the supply, ranged from several tenths of a second to several minutes. After the first explosion the phenomenon had appeared as a stationary glow. The flash occurred so suddenly that immediately before it not the slightest sign of a glow could be observed. This phenomena had

[1] Chariton and Walta, Z. Phys. 39, 547 (1926).

the character of an explosion[2] which took place at a certain pressure of the oxygen in the vessel.

At the same time the above-named authors measured with the McLeod gauge the increase of pressure from the moment of entry of the oxygen. It has been found that up until the moment of the flash the increase in the oxygen pressure in the vessel was uniform, as if indeed the entire amount of oxygen entering did not react at all with the phosphorus. The process took place in this way until the moment of the flash, which corresponds to a certain pressure P; from that moment there is no further increase in pressure, i.e. all of the additional oxygen entering is burnt. The authors named this pressure a critical or residual pressure and concluded that at a lower pressure the oxygen reacts either not at all or immeasurably slowly with the phosphorus vapour. As soon, however, as the pressure P is reached, the reaction immediately assumes a higher rate.

It has been found in additional experiments that if argon is present in the vessel besides the phosphorus vapour the residual pressure diminishes in such a way that the critical pressure P of the oxygen is lower the more argon is present. In addition, it has been established that P increases if the temperature of the phosphorus bulb increases.

[The next portion of this paper deals in detail with the experiments of Chariton and Walta, in the light of a criticism published by Bodenstein (*Z. für Physik* **41**, 548 (1927)). This portion is omitted here since it is mainly concerned with the significance of measurements made with the McLeod gauge. Semenoff concludes that the experiments of Chariton and Walta do establish the existence of an explosion limit, but agrees that the work is not quantitatively valid.]

Although, in my opinion, one can assume it to be proved by the work of Chariton and Walta that there is a residual pressure, I decided, taking into consideration all the inaccuracies of the method, to repeat their work without using a McLeod manometer and thus avoiding any possibility of the distillation of the phos-

[2] A. Strutt, *Proc. Roy. Soc.* A, **104**, 322 (1923), observed a similar phenomenon as also did Joubert, *Ann. Ecole Normale* **3** (1874).

E*

phorus from one part of the vessel to another, as Bodenstein recommends in his paper. In this way I obtained a series of new data on this remarkable phenomenon. I will now describe these new experiments, noting ahead that they are not finished and will be continued in the winter. In spite of that I decided to publish all the material obtained, for I considered it necessary to answer Bodenstein's criticism as soon as possible.

1. *Phenomena on Compression of the Gaseous Mixture in the Reaction Vessel and in the McLeod Gauge*

First of all, it is obvious that the McLeod gauge as a measuring instrument is inapplicable even when the trap with the liquid air is removed and all parts of the apparatus are at the same temperature. In this case the phosphorus distils into the McLeod gauge, and as the mercury is raised the glow takes place. The cause of this phenomenon is clear. On raising the mercury in the McLeod gauge we increase continuously the pressure of oxygen; as soon as it becomes higher than the residual pressure the oxidation of the phosphorus begins, and part of the oxygen is consumed (more or less, depending on how rapidly compression takes place). For this reason the pressure measured by the McLeod gauge bears no relation to the pressure of oxygen in the experimental vessel.

The use of the McLeod gauge was therefore rejected. In my opinion the occurrence of the explosions with compression of the mixture is a new and convincing proof of the existence of the residual pressure.

With Schalnikoff we carried out a series of observations on this phenomenon, by compressing the oxygen in a long tube of approximately 2 cm in diameter, containing phosphorus. First we introduced oxygen into the vessel at such a pressure that a glow over the whole tube could be seen. The supply of oxygen was then cut off, the glow died down, and we started to raise the mercury very gradually, with the aid of a micrometer screw, in order to compress the oxygen in the tube. We observed during this process a series of completely distinct explosions after each 2 mm rise of the mercury. As many as 30 such explosions, at regular intervals, were noted. When the mercury is raised to an arbitrary height X, then dropped abruptly, and then again raised, the first explosion occurs when it reaches a height beyond X. If the oxygen is subjected to less than the critical pressure, the first explosion does not take place until the pressure is increased to the critical pressure. The further course of the phenomenon remains the same; at each 2 mm rise an explosion is observed. If the presence of a residual pressure is postulated these experiments are easy to understand.

It is to be noted that a certain interval between the explosions indicates that the critical pressure of the flash and the residual pressure are not exactly equal, but that the latter is slightly lower.

Among other things it is to be noted that if a rather large quantity of oxygen is admitted to the vessel and an intense glow obtained, then after it is extinguished the level of the mercury must be raised not by 2, but by 5 mm in order to get the first explosion. To obtain subsequent explosions there must be further increases of 2 mm pressure. This means that the more intense is the explosion, i.e. the greater the initial oxygen pressure with respect to the critical pressure, the more oxygen is burnt, and the lower is the residual pressure. This also explains the peculiar phenomenon of the pulsating glow, which I will discuss later. To find out whether oxygen reacts at all at a pressure below the critical, we let the vessel stand for several hours, and then started the compression. It was found that after standing 6 to 7 hours it was necessary to compress the oxygen twice as much in order to obtain the first explosion. This obviously means that, in the course of the six hours, up to half the amount of the oxygen reacted without giving a glow. This result is in obvious contradiction to the experiment of Chariton and Walta, as well as to the result of the experiment described below. It seems that the gradual disappearance of the oxygen is connected with the large surface area of the mercury; either it catalyses the reaction or it adsorbs the oxygen.

2. *Admission of the Oxygen to the Vessel Containing the Phosphorus*

As the McLeod gauge had to be rejected as a measuring instrument a sulphuric acid manometer was used instead for recording the pressure–time curves. The arrangement is shown in Fig. 10.1. The 9 mm diameter tube A is evacuated to a pressure of 10^{-4} mm

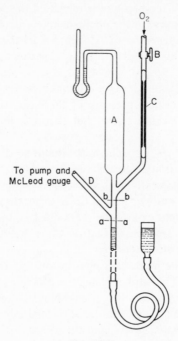

Fig. 10.1

by means of a series of mercury diffusion pumps; then a mercury valve isolates tube A from the pump and from the McLeod gauge (the mercury is set at the level b–b). Now the stop-cock B is turned on, and the oxygen (obtained by electrolysis of water and dried with P_2O_5) flows into the vessel A through the capillary tube C from the bulb (not shown in Fig. 10.1). The rate of the admission can be varied at will by changing the pressure of the oxygen in the containing vessel. The sulphuric acid manometer was read using a microscope. Eleven divisions of its eye-piece scale corresponded to 1 mm of displacement of the level in the arm.

Initially the pressure increases linearly with time; then, after a certain time-interval, the increase stops suddenly and the pressure remains constant. The transition time coincides with the instant of the explosion in the vessel; i.e. we obtained curves which are completely analogous to those found by Chariton and Walta in their work, although the accuracy of a sulphuric acid manometer is not especially high (the total pressure range amounts to 2·5 divisions of the scale); nevertheless quite good rate-of-admission curves can be obtained. The time of the transition of the meniscus was observed with a stop-watch, giving a satisfactory accuracy. However, the error could amount to 10 to 15 % of the value of the critical pressure. Twelve curves were obtained in all for different rates of admission; the result was always the same. The points where the curves changed their direction always coincided with the times of explosions. Three such curves are given in Fig. 10.2. The instant of explosion is marked with an arrow and was determined in a special experiment since darkness was necessary. These curves show that up to the occurrence of the explosion the oxygen does not react, while as soon as the glow occurs all the residual incoming oxygen is used up. In order to initiate the glow a certain critical pressure of oxygen is necessary.

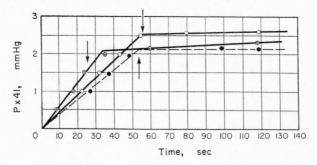

FIG. 10.2

Now I would like to discuss the phenomenon of the pulsating glow. At a certain intermediate flow-rate of oxygen the glow is not steady; it pulsates and the explosions occur at intervals of 1 second. With such a glow the meniscus of the manometer pulsates with the pulsation of the glow. (The amplitude is very small and difficult to measure; we estimated it as $0 \cdot 1$ of a division.) This phenomenon is easy to explain if the data on the difference between the pressure of the glow and the residual pressure are recalled. If the rate of flow through the capillary is sufficiently high to cause an intense explosion, then immediately after the explosion the pressure drops below that necessary to support the glow, and the glow is extinguished. The oxygen entering through the capillary induces a new explosion which in turn extinguishes the glow.

A similar phenomenon of pulsation, but with periods of several seconds and even sometimes of several tens of seconds, can be observed when the phosphorus distils into a bulb filled with oxygen. This phenomenon indicates that at a given pressure of oxygen a certain critical pressure of the phosphorus exists, which induces a reaction.

3. *Does a Reaction take place at a Pressure which is Lower than the Residual Pressure?*

It was further necessary to satisfy ourselves whether the observation of Chariton and Walta, that the oxygen can be in contact with the phosphorus for 48 hours without inducing an observable reaction, is correct.

For this purpose the following experiments have been carried out in the vessels shown in Fig. 10.1 (but provided with a stopcock instead of a mercury valve). The oxygen was pumped into a closed vessel until an explosion occurred (30 to 40 seconds). The supply of oxygen was then cut off and the vessel left under these conditions for a certain time t. After that more oxygen was admitted, and within 1 to 2 seconds after the penetration of the oxygen in the capillaries an explosion occurred. The interval t

was increased up to 24 hours and the same result was always obtained.

It follows that the residual pressure of oxygen in a vessel containing phosphorus does not decrease by any perceptible amount in 24 hours, which is in complete agreement with the experiments of Chariton and Walta. The gas which causes the residual pressure can hardly be some product of the oxidation of the phosphorus, because it does not condense in the liquid-air trap. This has been tested in the following way: a trap and a McLeod gauge were connected to the vessel; oxygen was pumped into the vessel until an explosion occurred and immediately the supply of the oxygen was cut off. If the mercury in the McLeod gauge was raised to a certain level a bright explosion occurred in the gauge. The decrease in volume in the McLeod gauge was approximately ten-fold and the residual pressure much higher than in the vessel, the dimensions of which were much larger than those of the McLeod bulb (see the following paragraphs). After that the trap was filled with liquid air and after 15 minutes the experiment repeated. The explosion in the McLeod gauge took place at the same level of the mercury. This proves that the gas in the vessel containing the phosphorus is not condensed in liquid air, and in all probability is oxygen. By these experiments we have proved the existence of a residual pressure of oxygen, below which it does not react perceptibly with phosphorus.

4. *The Dependence of the Critical Pressure on the Dimensions of the Vessel*

Schalnikoff accidentally discovered that the residual pressure in a bulb of diameter 26 cm amounted not to $0 \cdot 01$ mm but to $0 \cdot 0001$ mm. In this case the glow was confined to the bulb, the connecting tubes ($d = 6$ cm) remaining dark. The residual pressure was a function of the dimensions of the vessel. This dependence is so obvious that one is surprised that Chariton and Walta did not notice it. If the vessel has wide and narrow parts, then always only the widest part glows. The narrow parts glow only if a large quantity of the oxygen is pumped in so that the pressure therein is higher than the residual pressure.

I have studied these phenomena in the following experiments.

The four cylindrical vessels I, II, III and IV (Fig. 10.3) are all of the same length, 10 cm, and have inner diameters of $4 \cdot 6, 9, 20$

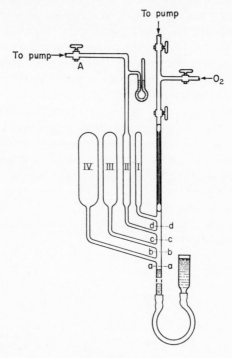

FIG. 10.3

and 31 mm; these are connected as shown in the figure. The time of the flash, i.e. the time which elapses between the admission of the oxygen and the explosion, is established with the valve A closed and with the mercury level at four different positions a, b, c, and d, which correspond to all four vessels being connected to the capillary, the first three, the first two, or only the first, respectively. Between each two experiments the valve A is opened, the mercury is lowered to the level a and the system evacuated to 10^{-5} mm. Each time the explosion occurs only the widest vessel glows. The pressure corresponding to the explosion in the widest vessel is proportional to $t/\Sigma V$, where t is the time between the admission of the oxygen and the beginning of the glow, ΣV the volume of the vessels and connecting tubing. Table 1 gives the time of the glow (t sec) for each of the four cases.

TABLE 1

Vessels connected to the capillaries	ΣV (cm³)	Diameter of the widest part (mm)	t_{obs} (sec)	t_{mean}	$\dfrac{t}{\Sigma V}$
I + II + III + IV	146	31	52, 52, 51, 49	51	0·35
I + II + III	59	20	53, 49, 55, 51, 49	51	0·86
I + II	22	9	45, 42, 48, 51, 49	47	2·14
I	4	4·6	25, 20, 25, 24, 22	23	5·75

The volumes of the vessels and connecting tubing were:

$$4 \text{ cm}^3 \quad 18 \text{ cm}^3 \quad 37 \text{ cm}^3 \quad 87 \text{ cm}^3$$

In the third experiment, when the glow occurred in the second vessel, the pressure was measured with a sulphuric acid manometer; it was found to correspond to 2·5 divisions of the eye-piece scale. Taking the specific gravity of the sulphuric acid as 1·82, the pressure P is calculated to be 61×10^{-3} mm Hg, with an accuracy of 10 to 15 per cent.

From that we compute the dependence of the pressure at which the explosion takes place on the tube diameter (Table 2).

TABLE 2

d (mm)	$P \times 10^3$ mm Hg
4·6	169
9	61
20	24
31	9

The measured values of the pressure may be represented by the equation

$$P_k \, d^{3/2} = \text{const.} \qquad \text{(I)}$$

as shown in Fig. 10.4.

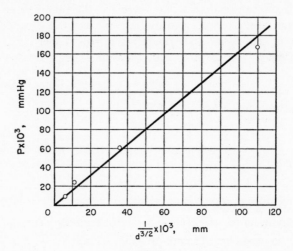

FIG. 10.4

5. *The Dependence of the Critical Pressure on the Volume and the Surface Area of the Vessel*

We have been convinced from a special experiment that for a cylindrical vessel the value of P_k is determined by the diameter and not by the volume. On the other hand for a sphere with the same diameter the pressure P_k is different. Two cylindrical vessels of 20 mm diameter were used, one (III) 51 cm, the other (II) 10 cm long. The volumes of these vessels with the connecting tubes and the part containing the phosphorus amounted to: $V_{\text{III}} = 133$ c.c., $V_{\text{II}} = 31$ c.c.; besides these a small sphere with the same diameter, 20 mm. (I) was added to the arrangement; its volume with the connecting tubing amounted to 8.9 c.c. In this experiment we studied three cases: (1) all three vessels connected by a capillary through which the oxygen was admitted; (2) only vessels II and I connected, and (3) only vessel I connected.

In the first case the explosion occurred simultaneously in III and II; I remained dark; in the second case the explosion occurred in I, and II remained dark.

TABLE 3

Vessels	ΣV (c.c.)	Length of the largest vessel (cm)	t_{obs}	t_{mean}	$\dfrac{t}{\Sigma V}$
I + II + III	173	51	183, 180, 187	183	1·06
I + II	40	10	45, 43, 45	44	1·10
I	8·9	Sphere	15, 12, 12	13	1·46

In the first case $t/\Sigma V = 1 \cdot 06$, in the second case $1 \cdot 1$ and in the third case $1 \cdot 46$, i.e. in the first and second cases the critical pressure is the same within the limits of the experimental error. The results are given in Table 3.

Now the ratio of the volume to the surface for a radius of 1 cm is 2 for a cylinder and 3 for a sphere; as one can see, the pressures at the explosion for the cylinder and the sphere are in approximately this ratio.

A similar experiment has been carried out with three glass cylinders of the same length (10 cm). Vessel III was empty ($d = 20$ mm); vessel II was of the same dimensions as III, but densely packed with pieces of glass tubing, each 10 mm long, with a diameter of 5·5 mm. Vessel I was empty, with a diameter of 5·5 mm.

The volumes were $V_I = 5$ c.c., $V_{II} = 18 \cdot 5$ c.c. and $V_{III} = 28$ c.c.

The results of the experiment are given in Table 4. The pressure of the explosion in an empty vessel was six times smaller than in that filled with the pieces of glass tubing, although the volume was only approximately twice as great.

TABLE 4

Vessels	ΣV (c.c.)	t_{obs}	t_{mean}	$\dfrac{t}{\Sigma V}$
I + II + III	52	19, 18, 20, 20	19	0·37
I + II	23·5	50, 56, 54, 58	55	2·34
I	5	15, 14, 17, 7	16	3·22

On the other hand, the explosion in the small vessel occurs at nearly the same pressure as in vessel II. Since the diameters are about the same for vessels II and I, the surface–volume ratio in vessel II is considerably greater than in I, so that the important role is played not so much by the surface–volume ratio as by the linear dimensions of the vessel, or better, the distance between the walls.

6. The Dependence of the Critical Pressure on the Temperature and the Pressure of the Phosphorus Vapour.

The experiment was carried out in the vessel illustrated in Fig. 10.5. The vessel B was placed in a water bath. The water could be heated by an electric heater E, which also served as a stirrer. Three small tubes came out of the vessel B; one was connected with the capillary C, through which the oxygen was supplied from a bulb not shown in Fig. 10.5; the second tube connected B with the diffusion pump and the McLeod gauge, so that the vessel could be isolated from the pump by a mercury valve by raising the mercury to the level b; the third tube led to the projection D which contained phosphorus. This projection was surrounded by a second water bath, F, which could also be heated by means of a heater. The temperature T_o of the water in the container A and the temperature T in the bath were read with a thermometer. The time interval t between the admission of the oxygen into the vessel B and the moment of the explosion was measured. Two series of experiments were carried out; during one the temperature of the bath, T, and therefore the pressure of the phosphorus vapour, was constant, while the temperature of the experimental bulb T_o was always higher than T. During the second series of experiments T_o was on the other hand kept constant, while the pressure of the vapour was varied within certain limits by changing T. Here, naturally, T was never higher than T_o.

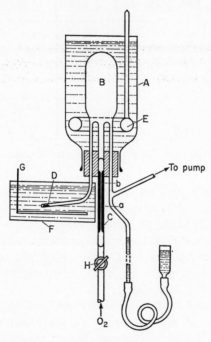

FIG. 10.5

In this manner the experiments were carried out always in a superheated vapour, so that the dependence of the critical pressure on the temperature and on the pressure of the phosphorus vapour could be studied separately.

First series of experiments The temperature of the projection with the phosphorus, $T = 17°C$; the temperature $T°$ of the vessel was varied. At different T_0 values the following times for the explosion were measured (Table 5).

TABLE 5 First series			TABLE 6 Second series†	
T_0	t in seconds	t_{mean}	T_0	t
17	44, 46	45	18	15
21	44, 44, 38, 44	43	19	17
25	46, 46, 45, 45	45	28	17
32	40, 44, 35	40	28	17
38	38, 37	38	31	18
			46	21
			46	19.5
			50	22

†The two series refer to different rates of supply.

From these figures we concluded that the critical pressure, which is proportional to t, is independent of the temperature at constant density of the phosphorus vapour.

Second series of experiments. The temperature of the vessel A, $T_0 = 39°C$; the temperature T of the bath F was varied.

The pressure P_{P_4} of the phosphorus is taken from the work by Dunoyer, *Technique du Vide*.

As can be seen from Fig. 10.6, the relationship between t and P_{P_4} or, since t is proportional to P_k, the relationship between P_k and P_{P_4}, can be represented by the formulae

$$t\sqrt{P_{P_4}} = \text{const.}; \quad P_k\sqrt{P_{P_4}} = \text{const.} \tag{II}$$

The results of the second series have been adjusted in such a way that the point $T = 16°$ has been brought to coincide with the corresponding point of the curve plotted for the first series; and from this has been determined the normalizing factor for all remaining figures of the second series.

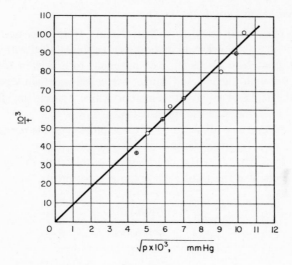

FIG. 10.6

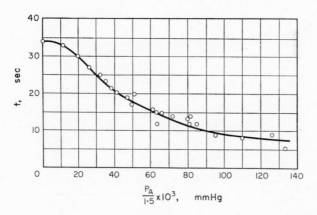

FIG. 10.7

7. *The Dependence of the Critical Pressure on the Amount of Argon added to the Phosphorus Vapour*

The cylindrical vessel ($d = 8$ to 9 mm), which has been provided with a sulphuric acid manometer, was next filled with argon (which contained up to 10% of nitrogen) to a certain pressure measured with the McLeod gauge; the vessel was then separated from the McLeod gauge with a mercury valve, and the time for the explosion, t, measured in the usual way.

The dependence of the time t of the explosion on the pressure of the argon at a given density of the phosphorus vapour ($16°C$, $P_P = 25 \times 10^{-3}$ mm Hg) can be seen on the curve in Fig. 10.7. The points give the results of separate experiments. In order to

TABLE 7

First Series

T_0	T	t in sec	t_{mean}	$P_{P_4}10^3$ mm Hg
39	35	11, 10·5	11	100
	26	13.5, 13·3	13·5	52
	25	15, 14, 15	15	50
	20	17, 18·5, 19	18	35
	12	27, 27·5	27	20

TABLE 8

Second Series

T_0	T	t in sec	t_{mean}	$P_{P_4}10^3$ mm Hg
41	16	18, 17, 15, 17	17	26
37	22	13, 12, 13	13	40
39	32	10, 10, 10	10	86
41	36	8·5, 8, 7	5·5	106

reduce the time data to partial pressures of the oxygen in 10^{-3} mm Hg it is necessary approximately to double the ordinates, for according to the readings of the sulphuric acid manometer the critical pressure in the absence of argon amounted to 65×10^{-3} mm Hg. The corresponding time t, however, as also seen in the graph, was 34 seconds. As Fig. 10.8 shows, the relation between the pressures of the components of the mixture and the critical pressure can be expressed[3] very accurately by the formula

$$P_k \left(1 + \frac{P_A}{P_k + P_{P_4}} \right) = \text{const.} \tag{III}$$

[3] After the present work was finished and written up Schalnikoff obtained new results for large spherical vessels. In his experiments the oxygen pressure was measured as follows. First the whole vessel was well evacuated. Then the rate of penetration of the oxygen through the capillaries was measured with a McLeod gauge for different pressures of oxygen in the supply vessel. The increase of the pressure with time was completely linear and reproducible for the same capillaries. After that the phosphorus was introduced into the vessel; this was evacuated to a pressure of approximately 10^{-5} mm Hg with a diffusion pump, and finally the oxygen was admitted. As the rate of admission was previously measured, the critical pressure could be calculated very accurately from the time elapsed up to the explosion. The following results have been obtained:

TABLE 9

Diameter of the vessel (d) in cm	Critical pressure P_k of the oxygen	$P_k d^2$
6	$5 \cdot 27 \times 10^{-3}$	190
13·4	$1 \cdot 16 \times 10^{-3}$	209
18·1	$0 \cdot 61 \times 10^{-3}$	200

As can be seen from the table, the results can be represented by the relation $P_k d^2 = \text{const.}$ Because of the low accuracy of the earlier measurement, which gave the relation $P_k d^{3/2} = \text{const.}$, we do not consider this difference to be very important. During these experiments another important property of the process was established, namely that in the oxidation of the phosphorus vapour the glow takes place not only on the inner wall, but throughout the whole vessel. One finds that if one looks along the diameter of the vessel the intensity of the glow is the highest in the centre and drops to zero at the walls. Two glass tubes were also connected diametrically to the glass sphere; on account of their small diameter they remained dark during the glow, but by looking along them the glow in the vessel could be observed in its full intensity.

F

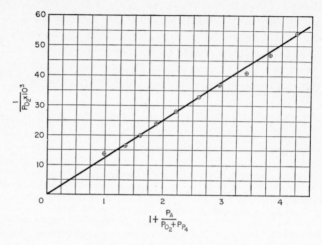

FIG. 10.8

Conclusions

We summarize our results in the following statements:

1. Phosphorus and oxygen react with one another either not at all, or the reaction is immeasurably slow, if the pressure of one of the components is below a certain critical pressure.

2. This critical pressure depends on the distance between the walls of the vessel, and it decreases rapidly with an increase of this distance. It seems that only with infinitely large dimensions of the vessel would the reaction take place at an arbitrarily low pressure of one of the reactants.

3. The addition of a neutral gas (argon) acts in the same way as the increase of the dimensions of the vessel: with an increase of the pressure of the argon the critical pressure of the oxygen decreases (at a constant phosphorus pressure).

4. The increase in the pressure of one of the components decreases the critical pressure of the other.

We do not attach great significance to the accuracy of the quantitative relationships obtained (laws I and II), but we believe that the qualitative results summarized in the above four points are absolutely correct.

From them it is possible to infer the following regarding the nature of the reaction between phosphorus and oxygen:

I. The surface of the vessel "poisons" the reaction, probably by absorption on its active centres. The longer is the time required by an active molecule to reach the wall of the vessel the smaller must be the contaminating effect of the wall. An increase in the distance between the walls of the vessel, or the addition of a neutral gas, makes it more difficult for the active molecules to reach the walls and therefore increases the probability that an active molecule will trigger a reaction before its activity is destroyed by the wall.

II. We see the reaction as a chain reaction: due to the heat motion certain activated initial centres will be created; each of them could cause a reaction which will produce by collisions of the second kind an increased quantity of the new centres of the same type; alternatively it could be destroyed through a collision with the wall, collisions of another kind, or by the radiation inside the vessel. If the probability of the deactivation is sufficiently high that the increase of the new centres, on a statistical average, will be not higher than unity, then a slow reaction takes place, whose rate is determined by the supply of the primary centres. The probability of the deactivation depends on the experimental conditions (pressures of the components, structure and pressure of the added gases, distance to the walls, etc.); however, when the gain of the new centres reaches and exceeds unity, the number of active centres increases with time and an explosive reaction occurs. Similar concepts have already been developed by Christiansen and Kramers.[4] In order to illustrate these relationships by an example, we wish to carry out calculations on the basis of some assumptions on the mechanism of the reaction between

[4] Christiansen and Kramers, *Z. Phys. Chem.* **104**, 451 (1923).

phosphorus and oxygen, but we do not assert that these assumptions correspond to reality. We will execute this computation only approximately, because an exact calculation is very difficult and would be unnecessary as a result of the uncertainty of the assumptions.

We assume that:

1. The active centres of the reaction are oxygen atoms (O).
2. These centres appear:

(a) As a result of spontaneous dissociation of the O_2 molecules. The number (n_0) originating in this way per unit time can be arbitrarily small.

(b) As a result of collisions of the second kind between the reaction products which have not yet had time to lose their energy, and the molecules of oxygen. Here we assume that by a collision of an excited molecule of the reaction product with an oxygen molecule O_2 the latter will be dissociated, but in a collision with a phosphorus molecule P_4 the excited molecule simply loses its energy. In order to obtain a clear picture we assume the following scheme for the reaction:

1. $O + P_4 = P_4O' \begin{cases} P_4O' + O_2 = P_4O + O + O \ (a) \\ P_4O' + P_4 = P_4O + P_4 \ (b) \end{cases}$

2. $P_4O + O_2 = P_4O_2 + O.$

3. $P_4O_2 + O_2 = P_4O_4 \begin{cases} P_4O_4' + O_2 = P_4O_4 + O + O \ (a) \\ P_4O_4' + P_4 = P_4O_4 + P_4 \ (b) \end{cases}$

4. $P_4O_4 + O_2 = P_4O_6.$

. . . .

6. $P_4O_8 + O_2 = P_4O_{10}' \begin{cases} P_4O_{10}' + O_2 = P_4O_{10} + O + O \ (a) \\ P_4O_{10}' + P_4 = P_4O_{10} + P_4 \ (b). \end{cases}$

Here the molecules which immediately after their formation have excess energy are marked with the sign $'$.

All members of the reaction proceed according to the same scheme, except for reaction 2, which is introduced in order to produce molecules with an even number of oxygen atoms. We can see from our scheme that an O-centre produces at the most eleven new centres, if each time the intermediate reactions follow reaction (a). In reality they can follow (b) also.

The probability of a collision with the O_2 is obviously equal to

$$a = \frac{P_{O_2}}{P_{P_4} + P_{O_2}},$$

where P_{O_2} is the pressure of the O_2 and P_{P_4} the pressure of the P_4.

A simple calculation shows that the average number of the new O-centres, which are produced by each reaction $P_4 + O = P_4O'$, amounts to $1 + 10a$. The one arises in the following way: each time reaction $P_4O + O_2 = P_4O_2 + O$ occurs one new O will be produced, while in the other five reactions two new O atoms originate with a probability a.

We now attempt to develop an equation for the rate of reaction, on the assumption that the reaction is in a stationary state, i.e. that the rate remains constant with time. It will be shown that a solution of this equation is possible under certain conditions; the boundary case, where the equation no longer holds, we consider to be the condition for the occurrence of the explosion, because there the rate, which has been considered constant in the construction of the equation, becomes infinitely large.

Let N be the number of the primary reactions $O + P_4 = P_4O'$ per unit time; then the number of the newly formed O-centres due to collisions of the second kind (in the chain process) will be equal to $N(1 + 10a)$. For the primary centres produced spontaneously, n_0 per unit time, we shall obtain on the whole $n_0 + (1 + 10a)N$ new centres per unit time.

On the other hand in unit time the O-centres disappear, partly because N primary reactions take place, in each of which one O-atom is destroyed, and partly because a certain number will be adsorbed by the wall. According to our scheme only these two possibilities exist for each O inside the vessel; hence, if we designate by A the probability that the O-atom on its way through the gas mixture will react with a P_4, then the probability that it will reach the wall will be equal to $1 - A$. Since now N is the number of the former processes per unit time, the number of the centres adsorbed by the wall is equal to $N(1 - A)/A$, and the total number of O-centres disappearing is equal to N/A.

If the whole process is in a stationary state, the equation

$$n_0 + (1 + 10\alpha)\,N - \frac{N}{A} = 0$$

must be satisfied; it says that the number of O-centres disappearing equals the number being formed. From that we obtain:

$$N = \frac{n_0 A}{1 - A(1 + 10\alpha)}.$$

Since the rate of the spontaneous decomposition of the oxygen molecules, n_0, is exceptionally small, the reaction rate N will also be very small, so that the following inequality applies:

$$A(1 + 10\alpha) < 1.$$

The smaller is n_0 the more sudden is the increase in N as the expression $A(1 + 10\alpha)$ approaches the critical value. At this value we obtain $N = \infty$, i.e. the reaction can not be stationary and an explosion occurs. We wish to express the value A in terms of the partial pressures of the components of the gas mixture and the dimensions of the vessel, and to do this we make some simplifying assumptions. If an O-centre has completed a path x in a straight line from the point of its origin and moves with a velocity c, then it needs according to a formula by Smoluchowski an average time

$$t = \frac{3\pi\,x^2}{4\lambda c}$$

during which it travels a path length ct and suffers a number of collisions $n = ct/\lambda$. If we assume that in the gas mixture argon is present besides O_2 and P_4, the number of collisions with P_4 is

$$\nu = n\,\frac{P_{P_4}}{P_{P_4} + P_{O_2} + P_{Ar}}.$$

If we assume that all molecules have the same diameter, λ will be inversely proportional to the total pressure, i.e.

$$\lambda = \frac{\lambda_0}{P_{P_4} + P_{O_2} + P_{Ar}}$$

where λ_0 is the free path length at a pressure of 1 mm Hg, the pressures being measured in millimetres of mercury. If one substitutes the expressions for n and λ in the formula for ν, one obtains

$$\nu = \frac{3\pi x^2}{4\lambda_0^2} (P_{P_4} + P_{O_2} + P_{Ar})P_{P_4}.$$

To a rough approximation we assume that the path which the O-centres must complete, in order to reach the wall of a cylindrical vessel, is on the average equal to $d/2$. Then we obtain, for the number of collisions with P_4 molecules which an O-centre undergoes on its way to the wall, the expression

$$\nu = \frac{3\pi}{16} \cdot \frac{d^2}{\lambda_0^2} (P_{P_4} + P_{O_2} + P_{Ar})P_{P_4},^{(5)}$$

This assumes that it completes its whole path without reacting with the P_4.

The probability that a reaction will take place at a single collision of the O with the P_4 we set equal to γ; the probability that during the whole path, with ν collisions, no reaction will take place is evidently $(1 - \gamma)^\nu$, and the probability that reaction takes place is

$$A = 1 - (1 - \gamma)^\nu = \nu\gamma - \frac{\nu(\nu - 1)}{1.2} \gamma^2 + \dots$$

If we assume that ν is considerably smaller than $1/\gamma$, then in the first approximation we can set

$$A = \nu\gamma.$$

Now we can write the complete equation for the critical boundary case:

[5] In our experiments this number varied between approximately 10 and 20.

$$1 = A(1 + 10a)$$

$$= \frac{3\pi}{16} \cdot \frac{d^2}{\lambda_0^2} (P_{P_4} + P_{O_2} + P_{Ar}) P_{P_4} \left(1 + \frac{10 P_{O_2}}{P_{O_2} + P_{P_4}}\right).$$

Here we note the following: if the vessel is infinitely large, then obviously the average probability of reaching the wall, $(1 - \gamma)^\nu$, is equal to zero, and $A = 1$. The critical condition is thus

$$1 + 10a = 1 \quad \text{or} \quad a = 0;$$

which means that an explosive reaction would take place at a very small pressure P_{O_2}.[6]

If a is not so small, the explosion takes place at a critical pressure of the O_2. For example, if

$$a = \frac{P_{O_2}}{P_{O_2} + P_{P_4}} = 0.5$$

it follows from the equations

$$A = \nu\gamma = \frac{1}{1 + 10a} = \frac{1}{6},$$

that the neglect of $(1/2)\nu^2\gamma^2$, as compared with $\nu\gamma$, will give an error of 8%.

The equation obtained can be transformed into

$$P_{P_4}(11 P_{O_2} + P_{P_4}) \left(1 + \frac{P_{Ar}}{P_{O_2} + P_{P_4}}\right) d^2 = \text{const.}$$

If P_{P_4} is considerably smaller than $11 P_{O_2}$ then as an approximation we obtain

$$P_{P_4} \cdot P_{O_2} \left(1 + \frac{P_{Ar}}{P_{O_2} + P_{P_4}}\right) d^2 = \text{const.}$$

---

[6] This result follows from the two assumptions of our mechanism, namely that each reaction chain produces at least one O (reaction 2), and that the O-centres will be deactivated only at the wall. We have neglected the recombination rate of the O-atoms to O_2-molecules on account of the extremely small concentration of O-centres in the steady state.

This relation represents more or less satisfactorily the general character of the relation found experimentally

$$P_{O_2} P_{P_4} \left(1 + \frac{P_{Ar}}{P_{O_2} + P_{P_4}} \right) d^{3/2} = \text{const.}$$

We thus believe that the essential features of our mechanism, namely, the initiation of the reaction by active centres, its chain character, and the poisoning effect of the wall, are realistic, and qualitatively explain the peculiarities of the phenomena observed. That the reaction is initiated by active centres and that the direct reaction $P_4 + O_2$ is impossible follows also from the experiments of Backer[7] who has established that in the complete absence of moisture the phosphorus does not react at all with the oxygen.

In conclusion I wish to express my thanks to Mr. M. W. Poljakow for his assistance in carrying out the experiments and to Prof. V. Bursian for the kind help with the theoretical part of this work.

[7] Backer, *Phil. Trans.* (1888).

THE THERMAL DECOMPOSITION OF ORGANIC COMPOUNDS FROM THE STANDPOINT OF FREE RADICALS. VI. THE MECHANISM OF SOME CHAIN REACTIONS†

F. O. RICE and K. F. HERZFELD

[Organic decompositions frequently have simple integral orders and it was natural to assume that they occur by simple molecular mechanisms; the ethane pyrolysis, for example, is usually first order, and was thought to occur by a simple splitting of the molecule into ethylene and hydrogen. The discovery by F. Paneth and W. Hofeditz (*Ber.* B, **62**, 1335 (1929)) of the free methyl radical led, however, to the suspicion that this and other radicals are involved in certain organic reactions, and in 1933 F. O. Rice and M. D. Dooley (*J. Amer. Chem. Soc.* **55**, 4245 (1933)) demonstrated the presence of free radicals in ethane that was undergoing decomposition. The question that then arose was: how is it possible for a reaction to occur by a complex mechanism and to exhibit simple kinetics? The paper that is reproduced below is of particular importance in that it shows that there are certain types of complex mechanisms that lead to simple kinetic behaviour.

Francis Owen Rice was born in England in 1890 but has lived in the United States since 1919; at the time this paper was written he was an Associate Professor at the John Hopkins University. He was later Chairman of the Chemistry Department at the Catholic University of America, and Chairman of the Chemistry Department of Georgetown University. He has made many pioneering contributions especially in the field of free radical reactions. Karl Ferdinand Herzfeld was born in Vienna in 1892 and has worked in the United States since 1923. At the time this paper was written he was Professor of Physics at the John

---

† A paper which appeared in the *Journal of the American Chemical Society*, **56**, 284 (1934).

Hopkins University, later becoming Chairman of the Department of Physics at the Catholic University of America. He has made many important contributions in theoretical physics, particularly in the fields of molecular structure, the solid state and ultrasonics.

The Rice–Herzfeld paper is here reproduced in its entirety. The reader should note that most of the numerical values quoted, e.g. for activation energies, now require substantial revision and that some of the mechanisms require modification; the general principles, however, still stand.]

ONE of us has recently proposed[1] a free radical mechanism for the pyrogenic decomposition of aliphatic organic compounds from which one can predict quantitatively the products formed on heating a wide variety of organic compounds.[2]

In order to substantiate this theory fully, it will be necessary to account also for the experimental fact that the decomposition of such substances as ethane, acetone, and dimethyl ether follows the equation of a unimolecular reaction, whereas the decomposition of acetaldehyde follows an equation[3] of an order between 1 and 2; we want to show here that these experimental results are consequences of the theory.

(A) Decomposition of Ethane

We will first write the chemical equations of the chain reactions, as postulated by the theory.

The velocity constant for the n-th reaction from left to right will be called k_n, from right to left, k'_n. Furthermore, the estimated heats of activation will be given since it is necessary to estimate the relative values of the k's.

---

[1] Rice, *J. Amer. Chem. Soc.* **53**, 1959 (1931); **55**, 3035 (1933).

[2] On the basis of this mechanism we would expect the experimental result that two organic compounds when mixed do not decompose independently of each other; see Heckert and Mack, *J. Amer. Chem. Soc.* **51**, 2706 (1929); Steacie, *ibid.* **54**, 1695 (1932); *J. Phys. Chem.*, **36**, 1562 (1932); Kassel, *J. Amer. Chem. Soc.* **54**, 3641 (1932); Steacie, *J. Chem. Phys.* **1**, 313 (1933).

[3] Kassel (*J. Phys. Chem.* **34**, 1166 (1930); *Annual Survey of American Chemistry*, 1932, p. 31) has mentioned these problems but was unable to solve them.

A. Reactions in the Decomposition of Ethane

Heat of activation

No.	Chemical equation	Velocity constant	E	E'
(1)	$C_2H_6 \rightleftarrows 2CH_3$	k_1, k'_1	80	8
(2)	$CH_3 + C_2H_6 \rightleftarrows CH_4 + CH_3CH_2$	k_2, k'_2	20	20
(3)	$CH_3CH_2 \rightleftarrows C_2H_4 + H$	k_3, k'_3	49	10
(4)	$H + C_2H_6 \rightleftarrows H_2 + CH_3CH_2$	k_4, k'_4	17	25
(5)	$H + H \rightleftarrows H_2$	$^1/_2 k_5, {}^1/_2 k'_5$	Triple collision	100
(6α)	$H + CH_3CH_2 \rightleftarrows C_2H_4 + H_2$	$k_6\alpha, k'_6\alpha'$	Small	60
(6β)	$H + CH_3CH_2 \rightleftarrows C_2H_6$	$k_6(1-\alpha), k'_6\beta$	Small	90
(7)	$H + CH_3 \rightleftarrows CH_4$	k_7, k'_7	Small	90
(8)	$CH_3 + CH_3CH_2 \rightleftarrows C_3H_8$	k_8, k'_8	8	80
(9)	$2CH_3CH_2 \rightleftarrows C_4H_{10}$	k_9, k'_9	8	80

The concentrations are denoted as follows

C_2H_6	CH_3	CH_3CH_2	H	CH_4
x_1	x_2	x_3	x_4	x_5

C_2H_4	H_2	C_3H_8	C_4H_{10}
x_6	x_7	x_8	x_9

If one writes the equilibrium conditions for (3) and (4) and multiplies the corresponding equations, one finds

$$\frac{x_6 x_7}{x_1} = \frac{k_3 k_4}{k'_3 k'_4} = K \qquad (3')$$
$$\text{(equil.)}$$

where K is the equilibrium constant. The equilibrium has been measured by Pease,[4] who finds as heat of reaction $31 \cdot 4$ kcal. Therefore $(3')$ gives

$$E_3 + E_4 - E'_3 - E'_4 = 31 \cdot 4. \qquad (4')$$

Similarly it follows from (6) that

$$E_\alpha + E_\beta - E_{1-\alpha} - E_{\alpha'} = 31. \qquad (6')$$

---

[4] Pease and Durgan, *J. Amer. Chem. Soc.* **52**, 1262 (1930).

Finally $E'_4 - E_4 = 103 - Q$, if 103 is the heat of dissociation of hydrogen and Q is the strength of the C—H bond in ethane (about 95); therefore

$$8 \leqq E'_4 - E_4 \leqq 12. \qquad (4'')$$

Furthermore, with $E_1 = 80$ and a strength of the C—C bond of 72–76, E'_1 must be between 4 and 8. We will confine our attention to the first stages of the reaction, so that all the back reactions may be neglected.[5] The proposed mechanism consists in the steps (1), (2), (3), (4) and (6).

We therefore derive first the kinetic consequences of these steps alone and later show that (5), (7), (8), (9) can be neglected with a proper choice of rate constants.

The concentrations of the intermediate radicals can be arrived at by assuming their amount constant in the steady state (after an immeasurably short period).

$$dx_2/dt = k_1 x_1 - k_2 x_1 x_2 = 0, \qquad (10)$$

$$dx_3/dt = k_2 x_1 x_2 - k_3 x_3 + k_4 x_1 x_4 - k_6 x_3 x_4 = 0, \qquad (11)$$

$$dx_4/dt = k_3 x_3 - k_4 x_1 x_4 - k_6 x_3 x_4 = 0. \qquad (12)$$

From these equations[6]

$$x_2 = k_1/k_2, \text{ and} \qquad (13)$$

$$x_3 = x_1 \sqrt{(k_1 k_4/2k_6 k_3)}, \qquad (14)$$

$$x_4 = \sqrt{(k_1 k_3/2k_4 k_6)}. \qquad (15)^{[7]}$$

The number of hydrogen atoms, x_4, is then independent of x_1, the amount of ethane (at moderate pressures). The reason for this

[5] Of course in equilibrium the rate of each reaction is equal to that of its own back reaction (principle of detailed balancing).

[6] At low pressures (10) would take the form $k_1 x_1 - k'_1 x_2^2 = 0$. At high pressures $k'_1 x_2^2 : k_2 x_2 = k'/k_1 : k_2^2$ or $10^5 - (E_1 + E'_1 - 2E_2)/2 \cdot 3\ RT$ which is small compared with 1 if $E_2 < 32$.

[7] The exact equation is

$$\frac{k_1 k_4}{2k_3 k_6}\left(1 - k_8 - \frac{k_1}{k_2}\right) = x_4^2 \left\{1 + \frac{k_5}{2k_6}\frac{k_3 - k_8 k_1/k_2}{k_4}\frac{1}{x_1} - \right.$$

$$\left.\frac{k_9}{2k_6}\frac{k_4}{k_3 - k_8\,k_1/k_2}\,x_1\right\} + \left(\frac{k_8}{k_9} + \frac{k_7}{k_6}\frac{k_3 - k_8\,k_1/k_2}{k_4}\frac{1}{x_1}\right)\frac{k_1}{k_2}\,x_4.$$

is that the production of H through (3) as well as its destruction through (6) are both proportional to x_1.

We have taken into account the fact that the rates of formation of ethylene and methane are k_3x_3 and k_1x_1, respectively; since the methane yield is small, we must have

$$k_3x_3 \gg k_1x_1 \quad \text{or} \quad \sqrt{(k_1k_3k_4/2k_6)} > k_1. \tag{16}$$

Furthermore, in order to exclude the chain-destroying reactions (5), (7), (8) and (9) we have assumed that

$$\tfrac{1}{2}k_5x_4{}^2 < k_6x_3x_4 \quad \text{or} \quad \frac{k_5}{2k_6}\frac{k_3}{k_4} < x_1, \tag{17}$$

$$\tfrac{1}{2}k_9x_3{}^2 < k_6x_3x_4 \quad \text{or} \quad \frac{k_9}{2k_6}\frac{k_4}{k_3}x_1 < 1, \tag{18}$$

$$k_7x_2x_4 < k_6x_3x_4 \quad \text{or} \quad \frac{k_7k_1}{k_2} < \sqrt{\left(\frac{k_1k_4k_6}{2k_3}\right)}, \tag{19}$$

$$k_8x_2x_3 < k_6x_3x_4 \quad \text{or} \quad \frac{k_8k_1}{k_2} < \sqrt{\left(\frac{k_1k_3k_6}{2k_4}\right)}x_1. \tag{20}$$

Before discussing these equations, we will write the over-all equation for the decomposition of ethane:

From (13), (14) and (15)

$$\frac{dx_1}{dt} = -x_1\left[\frac{3}{2}k_1\left(1 - \frac{k_3(1-a)}{3k_4}\right) + \sqrt{\left(\frac{k_1k_3k_4}{2k_6}\right)}\right]. \tag{21}$$

Of this, $-\tfrac{1}{2}k_1x_1$ is the primary reaction, and

$$2 + a + 2\sqrt{\left(\frac{k_3k_4}{2k_1k_6}\right)} \sim 2e^{(E_1-E_3-E_4)/2RT}$$

is the length of the chain.

Marek and McClure[8] have measured the heat of activation and have found that

$$\tfrac{1}{2}(E_1 + E_3 + E_4 - E_6) = 73 \text{ kcal} \quad \text{or} \quad E_3 + E_4 = 66. \tag{22}$$

In the discussion of reactions (16)–(20), we remark that Marek's measurements[8] extend over temperatures from 873–973° abs. and x_1 from $\tfrac{1}{3}$ to $\tfrac{1}{8}$; we measure x in volume concentrations relative

[8] Marek and W. B. McClure, *Ind. Eng. Chem.* **23**, 878 (1931).

to normal pressure and temperature. If we assume that the accuracy of the first order is assured within 10%, the inequality sign in (17) and (18) means that the left side is not larger than one-fifth of the right (see footnote 7, quadratic equation).

We assume that all unimolecular reactions have rate constants $10^{14} e^{-E/RT}$ sec^{-1}, while for all bimolecular reactions the constant is $10^9 e^{-E/RT}$, with the following exceptions: the reunion of 2H needs a triple collision and has therefore a rate of about $k_5 x_4^2 = 10^6 x_4^2$; to satisfy both (17) and (18), it is necessary that $k_9/k_6 = 10^{-3}$.

If $E_9 = 8$ kcal, there must be an additional steric or probability factor of 10^{-1} in k_9. If the strength of the C—C bond were 76 instead of 72, E_9 would equal 4 and a factor of 10^{-2} would be necessary. As there is no reason to single out reaction (9), we will assume this factor $\frac{1}{10}$ and $E = 8$ kcal for all reactions involving the reunion of radicals.

Then (16) is fulfilled automatically, since from (22), equations (17) and (18) are equivalent to $k_9/1 \cdot 6k_6 = k_3/k_4 = 50$ or $32 \geq E_3 - E_4 \geq 28$ (24). The values chosen in Table A comply with this inequality; they are selected in view of equation (27′) of the back reaction and condition (4″). (19) and (20) are fulfilled with a wide margin.

The final equation for the decomposition of ethane is then

$$\frac{dx_1}{dt} = -kx_1; \log k = \tfrac{1}{2} \log \frac{k_1 k_3 k_4}{2k_6} = 13 \cdot 7 - \frac{73,000}{2 \cdot 3RT} \qquad (22')$$

compared with Marek and McClure's empirical value

$$\log k = 15 \cdot 12 - \frac{73,200}{2 \cdot 3RT}. \qquad (22'')$$

The length of the chain is

$$2e^{(80-17-49)/2RT} \sim 100$$

which is in reasonably good agreement with the fact that 2% of methane has been found among the decomposition products (rate of formation $k_2 x_1 x_2 = k_1 x_1$). The chain length is rather sensitive to changes in $E_3 + E_4$.

(A') The Formation of Ethane from Hydrogen and Ethylene

This reaction has been investigated by Pease.[9] The mechanism follows without any new assumption directly from our previous investigation.

The mechanism involves $(6a')$ as production of the chain-carriers H and CH_3CH_2, and then the chain $(3')$ and $(4')$.

As equations for the condition of constancy of the carriers we have for H:

$$k'_6 a' x_6 x_7 - k_6 x_3 x_4 + k_3 x_3 - k'_3 x_4 x_6 + k'_4 x_3 x_7 - k_5 x_4{}^2 = 0 \quad (23)$$

and for CH_3CH_2:

$$k'_6 a' x_6 x_7 - k_6 x_3 x_4 - k_3 x_3 + k'_3 x_4 x_6 - k'_4 x_3 x_7 = 0. \quad (24)$$

The main reaction, disappearance of ethylene, is given by

$$-\frac{dx_6}{dt} = k'_6 a' x_6 x_4 + k'_3 x_4 x_6 - k_3 x_3 - k_6 a x_3 x_4. \quad (25)$$

In (23) $k_6' x_4{}^2$ can be neglected in comparison with $k'_3 x_4 x_6$, as is shown by inserting (26). (23) and (24) lead to

$$x_3{}^2 = x_6{}^2 \frac{k'_3 k'_6 a' x_7}{k'_4 x_7 + k_3}, \quad (26)$$

or

$$-\frac{dx_6}{dt} = k_6(1-a)x_3 x_4 + k'_4 x_3 x_7 = k'_6(1-a)x_6 x_7 +$$
$$x_6 x_7 \sqrt{\left(\frac{k'_3 k'_6 a' k'_4{}^2 x_7}{k'_4 x_7 + k_3}\right)}. \quad (27)$$

Pease's investigations show the reaction to be unimolecular within the range of 773–825°K. and down to $x_7 \sim \frac{1}{6}$. Again taking 10% as the limit of accuracy, this means that

$$k_3/k'_4 \times 6 \leqq 1/5, \quad \text{or} \quad E_3 - E'_4 \geqq 24 \quad (27')$$

which is fulfilled by the values of the heats of activation which we have selected.

[9] R. N. Pease, *J. Amer. Chem. Soc.* **54**, 1876 (1932).

We find therefore for the resultant reaction

$$\sqrt{\left(\frac{k'_3 k'_6 a'}{k_6}\right)} k'_4 \times x_6 x_7 \tag{28}$$

with a resultant heat of activation

$$\tfrac{1}{2}(E'_3 + E'_4 + E'_\alpha) = \tfrac{1}{2}(35 + 60) = 47 \cdot 5 \text{ kcal} \tag{28'}$$

which value is in good agreement with the experimental value of
Pease, $43 \cdot 5$.

If this is the predominant mechanism rather than the direct
reaction, then the agreement of the velocities as measured by
Marek (forward reaction) and by Pease (back reaction) with the
equilibrium is somewhat fortuitous.

The sum of the apparent heats of activation of the forward and
back reactions according to our calculations is

$$\tfrac{1}{2}(E_1 + E_3 + E_4 - E'_3 - E'_4 - E'_\alpha)$$

or according to (4')

$$\tfrac{1}{2}(E_1 - E'_\alpha + Q)$$

or according to (6').

$$Q + \tfrac{1}{2}(E_1 + E_{1-\alpha} - E_\alpha - E_\beta)$$

and this is Q only if $E_1 - E'_\beta - E_\alpha + E_{1-\alpha} = 0$.

(B) Decomposition of Acetone

Again numbering the equations and substances as in the case of
ethane, we have

Reaction

(1)	$CH_3COCH_3 \rightleftarrows CH_3 + CH_3CO$
(2)	$CH_3CO \rightleftarrows CH_3 + CO$
(3)	$CH_3 + CH_3COCH_3 \rightleftarrows CH_4 + CH_3COCH_2$
(4)	$CH_3COCH_2 \rightleftarrows CH_2 = CO + CH_3$
(5)	$CH_3 + CH_3 \rightleftarrows C_2H_6$
(6)	$CH_3 + CH_3COCH_2 \rightleftarrows CH_3COC_2H_5$
(7)	$2CH_3COCH_2 \rightleftarrows (CH_3COCH_2)_2$

	Reaction constant	Heat of activation	
(1)	k_1, k'_1	$E_1 = 70$	
(2)	k_2, k'_2	$E_2 = 10$	
(3)	k_3, k'_3	$E_3 = 15$	
(4)	k_4, k'_4	$E_4 = 48$	
(5)	k_5, k'_5	$E_5 = 8$	
(6)	k_6, k'_6	$E_6 = 8$	
(7)	k_7, k'_7	$E_7 = 8$	$E'_7 = 45$

As before, we number substances, denoting concentrations as follows:

CH_3OCH_3		CH_3	CH_3CO	CH_3COCH_2
x_1		x_2	x_3	x_4
CH_2CO	CO	CH_4	C_2H_6	$CH_3COCH_2CH_3$
x_5	x_6	x_7	x_8	x_9

Proceeding in a manner similar to A we arrive at the following equations for the intermediate products as condition for the steady state

$$\frac{dx_2}{dt} = k_1x_1 + k_2x_3 - k_3x_1x_2 + k_4x_4 - k_6x_2x_4 = 0, \qquad (8)$$

$$\frac{dx_3}{dt} = k_1x_1 - k_2x_3 = 0, \qquad (9)$$

$$\frac{dx_4}{dt} = k_3x_2x_1 - k_4x_4 - k_6x_2x_4 = 0, \qquad (10)$$

which lead to the following results

$$x_3 = (k_1/k_2)x_1, \qquad (9')$$

$$x_2 = \sqrt{(k_1k_4/k_6k_3)}, \qquad (11)$$

$$x_4 = x_1 \sqrt{(k_1k_3/k_6k_4)}. \qquad (12)$$

The concentration of CH_3COCH_2, the main carrier of the chain, is proportional to x_1 (concentration of CH_3COCH_3) because it is produced by a collision of methyl radicals with acetone (the concentration of methyl is constant, since it is produced by the dissociation of acetone and disappears by collision with CH_3COCH_2, which has a concentration proportional to that of acetone).

The following inequalities have been assumed

$$k_6 x_2 x_4 \ll k_4 x_4 \text{ or } k_1 k_6 / k_4 k_3 \ll 1, \tag{13}$$

$$k_5 x_2{}^2 \ll k_1 x_1 \text{ or } k_5 k_4 / k_6 k_3 \ll x_1. \tag{14}$$

Equation (13) leads to

$$E_3 + E_4 - 78 \ll 0 \tag{13'}$$

and is easily fulfilled. (14) means that

$$10^5 \, e^{(E_6 - E_5 + E_3 - E_4)/RT} \ll x_1. \tag{14'}$$

Within the range of the experiments ($T = 800$ to $900°$ abs., $x_1 \geqq \frac{1}{24}$) and with the accuracy we have assumed before,

$$E_4 - E_3 \geqq 26.$$

We can neglect reaction (9), since after a few seconds a steady state of equality between reactions (9) and (9′) is reached.[10]

One finds then for the decomposition of the acetone[11]

$$dx_1/dt = k_1 x_1 - k_3 x_2 x_1 \tag{15}$$

---

[10] One easily finds $x_{10} = k_7 x_4{}^2/k'_7 \, (1 - e^{-k'_7 t})$; x_{10} has reached $1 - 10^{-5}$ of its final value after a time $11 \cdot 5/k'_7 = 10^{-13} \, e^{45/RT}$, which is only a few seconds at the utmost. At that time the excess of CH_3COCH_2 combining with itself over the reproduction of this radical through the dissociation of $(CH_3COCH_2)_2$ is $10^{-5} \, k_7 x_4{}^2 = 10^{-5} k_7 \, (k_1 k_3 / k_6 k_4 x)$. This is small compared with $k_1 x_1$. The stationary value of x_{10} is $k_7 x_4{}^2/k'_7 = 10^{-5} \, e^{-16/RT}$.

[11] In the decomposition of both acetone and dimethyl ether there are two chain carriers, the methyl group and a heavy carrier. The chain could be terminated in three ways: by recombination of two methyl groups, by recombination of two heavy groups, or by combination of one methyl and one heavy group. The latter mechanism is necessary to give the right order of reaction. It is predominant over the recombination of two methyl groups, because with the assumed heats of activation there are more heavy radicals than methyl groups present; the recombination of the two heavy radicals is relatively insignificant because of the chemical instability of the bond which is formed, thus permitting the resulting product to be decomposed very quickly into its components. We have not assigned a large activation energy to the recombination of methyl radicals (see Heitler and Schuchowitzki, *Phys. Zeit. der Sowjetunion*, **3** (1933)), because our experimental work indicates that this is contrary to the experiments.

or leaving out unimportant terms

$$dx_1/dt = -kx_1, \text{ with } k = \sqrt{(k_1 k_3 k_4/k_6)} \qquad (15')$$

$$\log k = 14 \cdot 5 - (62,500/2 \cdot 3RT) \qquad (15'')$$

while Hinshelwood and Hutchinson[12] give $15 - (68,500/2 \cdot 3RT)$. The chain length is given by

$$\sqrt{\left(\frac{k_3 k_4}{k_1 k_6}\right)} \sim 3 \, e^{15/2RT} \sim 300.$$

Thus the formula for the chain length is, except for a factor 2, the same as that for ethane, the numerical difference being due to the difference in the heats of activation.

(C) Decomposition of Dimethyl Ether

This reaction has been investigated by Hinshelwood.[13] The mechanism according to the theory of free radicals is given below.* For an intermediate stage two ways of reaction are possible, namely, (2), or (3) + (4). It will be seen later that the results of the calculations are the same for both alternatives.

We again give indices to the different substances

CH_3OCH_3	CH_3	CH_3O	H	CH_2OCH_3	CH_4
1	2	3	4	5	6

HCO	H_2	C_2H_6	CH_3OH	$CH_3CH_2OCH_3$
7	8	9	10	11

and use x with the corresponding subscript to denote concentration.

Only CH_3 (2) and CH_2OCH_3 (5) are involved as intermediaries in the chain. Accordingly CH_3O (3), and H (4) in the second mechanism, are present in much smaller quantities. Therefore, we are going to neglect beforehand the reactions in which (2) or (5) disappear by collision with (3) or (4), that is, reactions (10) to (14).

---

[12] Hinshelwood and Hutchinson, *Proc. Roy. Soc.* A (London), **111**, 245 (1926).

[13] Hinshelwood and Askey, *ibid.* A, **115**, 215 (1927).

* [On p. 165].

Decomposition of Dimethyl Ether

No.	Chemical equation	Reaction constant	E	E'
(1)	$CH_3OCH_3 \rightleftarrows CH_3 + CH_3O$	k_1, k'_1	80	
(2)	$CH_3O + CH_3OCH_3 \rightleftarrows CH_3OH + CH_2OCH_3$, or	k_2, k'_2	15–25	
(3)	$CH_3O \rightleftarrows HCHO + H$, and	k_3, k'_3	20–45	
(4)	$H + CH_3OCH_3 \rightleftarrows H_2 + CH_2OCH_3$	k_4, k'_4	10–15	
(5)	$CH_3 + CH_3OCH_3 \rightleftarrows CH_4 + CH_2OCH_3$	k_5, k'_5	15	
(6)	$CH_2OCH_3 \rightleftarrows CH_3 + HCHO$	k_6, k'_6	38	
(7)	$2CH_3 \rightleftarrows C_2H_6$	k_7, k'_7	8	80
(8)	$CH_3 + CH_2OCH_3 \rightleftarrows C_2H_5OCH_3$	k_8, k'_8	8	45
(9)	$2CH_2OCH_3 \rightleftarrows (CH_3OCH_2)_2$	k_9	8	45
(10)	$2CH_3O \rightleftarrows (CH_3O)_2$ unstable		8	
(11)	$CH_3O + CH_2OCH_3 \rightleftarrows CH_3OCH_2OCH_3$	k_{11}		
(12)	$H + CH_3 \rightleftarrows CH_4$	k_{12}		
(13)	$H + CH_3O \rightleftarrows CH_3OH$	k_{13}		
(14)	$H + CH_2OCH_3 \rightleftarrows CH_3OCH_3$	k_{14}		

(5), (6) } chain

We start with the consideration of the first mechanism and find for the three intermediate compounds

$$CH_3: \ dx_2/dt = k_1 x_1 - k_5 x_1 x_2 + k_6 x_5 - k_7 x_2^2 - k_8 x_2 x_5 = 0, \quad (16)$$

$$CH_3O: \ dx_3/dt = k_1 x_1 - k_2 x_1 x_3 - k'_1 x_2 x_3 = 0, \quad (17)$$

$$CH_2OCH_3: \ dx_5/dt = k_2 x_1 x_3 + k_5 x_1 x_2 - k_6 x_5 - k_8 x_2 x_5 - k_9 x_5^2 = 0. \quad (18)$$

From these equations

$$x_3 = k_1/k_2, \quad (19)$$

$$x_2 = \sqrt{(k_1 k_6 / k_8 k_5)}, \quad (20)^{(14)}$$

$$x_5 = x_1 \sqrt{(k_1 k_5 / k_8 k_6)}. \quad (21)$$

The following reactions have been neglected

$$k'_1 x_2 x_3 \ll k_2 x_1 x_3 \quad \text{or} \quad k'_1 \sqrt{(k_1 k_6 / k_8 k_5)} \ll k_2 x_1, \quad (22)$$

$$k_7 x_2^2 \ll 2 k_8 x_2 x_5 \quad \text{or} \quad k_7 k_6 / 2 k_8 k_5 \ll x_1. \quad (23)$$

(22) is automatically fulfilled, and (23) gives

$$(10^5/2) e^{(E_5 - E_6)/RT} \leqq x_1/5 \quad (23')$$

with the same accuracy as before; or with $x_1 \sim \frac{1}{6}$, $E_5 - E_6 \geqq 23$.

The rate of decomposition of the ether is given by $dx_1/dt = -2k_1 x_1 - k_5 x_1 x_2$ or

$$\log k = \tfrac{1}{2} \log \frac{k_1 k_5 k_6}{k_8} = 14 \cdot 5 - \frac{62{,}500}{2 \cdot 3RT}. \quad (24)$$

Hinshelwood gives $13 \cdot 2 - (58{,}500/2 \cdot 3RT)$.

Methane is formed at the rate $k_5 x_2 x_1$ (see dimethyl ether). Formaldehyde, HCHO, at the same rate, since $k_6 x_5 = k_5 x_2 x_1$.

CH_3OH: $k_2 x_3 x_1 = k_1 x_1$, which is equal to the primary reaction.

$CH_3CH_2OCH_3$: $k_8 x_2 x_5 = k_1 x_1$, which is again equal to the primary reaction.

C_2H_6: $k_7 x_2^2 = k_7 k_1 k_6 / k_5 k_8$, which is less than the primary reaction.

---

[14] The exact formula for x_2 is $x_2^2 = \dfrac{k_1 k_6}{k_8 k_5} \dfrac{1}{1 + (k_7 k_6/2k_8 k_5)\,(1/x_1)}.$

In deriving it, $k_9 x_5^2$ has not been considered for the same reasons as the analogous reaction in the case of acetone.

In the low-pressure region this situation is reversed; the formation of ethane is then equal to the primary reaction and responsible for the breaking of the chain, while the formation of ethyl methyl ether, which terminates the chain in the high-pressure region, is now proceeding at the rate $(k_8 k_5 / k_7 k_6) 2 k_1 x_1^2$, which is slower than that of the primary reaction.

In the low pressure range, we have for the expression $k_5 x_2 x_1$, which governs the disappearance of the ether and the production of methane and formaldehyde

$$k_5 \sqrt{\frac{2k_1}{k_7}} \, x_1^{3/2} \sim 10^{12} \, e^{-(2E_5 + E_1 - E'_7)/2RT} \, x_1^{3/2} \qquad (25)$$

or $$\ln k' = 27 \cdot 6 - (51{,}000/RT). \qquad (25')$$

In the high-pressure range, the chain length is given by

$$\sqrt{\left(\frac{k_6 k_5}{k_8 k_1}\right)} \sim 3 e^{-(E_6 + E_5 - E_1 - E_8)/2RT} \sim 200{,}000 \ (\text{at } 800° \text{ abs.}) \quad (26)$$

the formula being similar to that arrived at in the studies of the decomposition of ethane and acetone.

We have now to discuss the alternative mechanism, given by (3) and (4), instead of (2). One finds instead of (19) that $x_3 = k_1/k_3$, but (20) and (21) are unchanged. Therefore this mechanism leads to approximately the same results as the other except for the fact that instead of methyl alcohol, hydrogen is formed at the rate $k_4 x_1 x_4$ which is smaller than $k_1 x_1$.

Kassel and also Steacie[15] have shown that in a mixture of dimethyl ether and ethyl ether each compound decomposes faster than would correspond to its partial pressure. These authors explain this through a so-called "cross activation". However, this phenomenon follows directly from the theory of free radicals, as each of the ethers supplies the radicals which enter into the chain reactions of both.

[15] Kassel, *J. Amer. Chem. Soc.* **54**, 3641 (1932); Steacie, *J. Chem. Phys.* **1**, 313 (1933).

(D) Decomposition of Acetaldehyde

This reaction has been investigated by Hinshelwood and Hutchinson,[12] Fletcher and Hinshelwood[16] and by Kassel.[17] The latter found the initial rate to be proportional to the $\frac{5}{3}$ power of the concentration. From the ratio of the time it takes to decompose the aldehyde by $\frac{1}{2}$ to the time it takes to decompose it to $\frac{1}{3}$, Fletcher and Hinshelwood conclude that a succession of three different second-order reactions occur in the pressure range from $0 \cdot 2$ to 1000 mm. If one plots, however, the logarithm of the initial velocity against the logarithm of the pressure as did Kassel one gets very accurately the $1 \cdot 5$ order reaction in the range between 1000 and 100 mm or from 400 to 40 mm (these two ranges are for runs at different temperatures).

It will be shown here that a rate proportional to the $1 \cdot 5$ power of the concentration follows directly from the theory of free radicals, as does also the composition of the products, as well as the heat of activation of the reaction.

The theory as far as it is developed here does not predict what happens in the case of such large percentage decompositions as are considered in Fletcher and Hinshelwood's calculation.

Reaction Mechanism

No.	Chemical equation	Velocity constant	Heat of activation E	E'
(1)	$CH_3CHO \rightleftarrows CH_3 + HCO$	k_1, k'_1	70	8
(2)	$HCO \rightleftarrows CO + H$	k_2, k'_2		
(3)	$H + CH_3CHO \rightleftarrows H_2 + CH_3CO$	k_3, k'_3		
(4) ⎫ chain	$CH_3CO \rightleftarrows CH_3 + CO$	k_4, k'_4	10	
(5) ⎭	$CH_3 + CH_3CHO \rightleftarrows CH_4 + CH_3CO$	k_5, k'_5	15	
(6)	$2CH_3 \rightleftarrows C_2H_6$	k_6, k'_6	8	
(7)	$CH_3 + CH_3CO \rightleftarrows CH_3COCH_3$	k_7, k'_7	8	
(8)	$2CH_3CO \rightleftarrows CH_3COCOCH_3$	k_8, k'_8	8	45

[16] Fletcher and Hinshelwood, *Proc. Roy. Soc.* A (London), **141**, 41 (1933).
[17] Kassel, *J. Phys. Chem.* **34**, 1166 (1930).

We again number the concentrations:

CH₃CHO	CH₃	HCO	H	CH₃CO
x_1	x_2	x_3	x_4	x_5

The difference in reaction order of acetaldehyde and the other compounds discussed in this paper is due to the fact that the two carriers of the chain, CH_3 and CH_3CO, are present in quite different amounts (because of the particular values of the heats of activation). Therefore, the more abundant of the two, CH_3, disappears finally through recombination, according to (6), instead of through combination with CH_3CO (according to 7), as happens in the other decompositions considered in this paper.

The steady state of the two intermediate substances not occurring in the chain is given by

$$\text{for HCO: } k_1 x_1 - k_2 x_3 = 0, \tag{9}$$

$$\text{for H: } \quad k_2 x_3 - k_3 x_4 x_1 = 0. \tag{10}$$

It follows immediately that

$$x_3 = k_1 x_1 / k_2, \tag{9'}$$

$$x_4 = \frac{k_2 x_3}{k_3 x_1} = \frac{k_1}{k_3}. \tag{10'}$$

For the two carriers of the chain we find:

$$\text{for CH}_3 \text{: } k_1 x_1 + k_4 x_5 - k_5 x_2 x_1 - k_6 x_2^2 - k_7 x_2 x_5 = 0, \tag{11}$$

$$\text{for CH}_3\text{CO: } k_3 x_1 x_4 - k_4 x_5 + k_5 x_2 x_1 - k_7 x_2 x_5 = 0. \tag{12}$$

These equations lead to

$$x_2 = \sqrt{(k_1/k_6)} \sqrt{x_1}, \tag{13}$$

$$x_5 = \sqrt{(k_5/k_6)} x_1^{3/2}. \tag{14}$$

For the final reaction, the disappearance of acetaldehyde, one gets
$$-dx_1/dt = 2k_1 x_1 + k_3 x_1 x_4 + k_5 x_2 x_1 = 2k_1 x_1 + k_5 \sqrt{(k_1 k_6)} x_1^{3/2}. \tag{15}$$

The experimental value for the heat of activation is 45·5 kcal, the theoretical value is

$$\tfrac{1}{2}(E_1 - E_6 + E_5) = 46.$$

The constant factor is in our units (1 mole in 22,400 c.c. at 0°) about 10^{12}. Kassel's value, divided by $(22,000)^{2/3}$ is 6×10^{10}.

We have assumed $k_6 x_2^2 \ll k_4 x_5$, $k_5/k_4 \ll k_6/k_7$, both of which are amply fulfilled. We are grateful to Dr. L. Kassel for constructive criticism in connection with this paper.

Summary

A discussion of the reaction mechanism postulated by the theory of free radicals shows that the decompositions of ethane, acetone and dimethyl ether must be of the first order, the dissociation of acetaldehyde of the 1·5th order, while the formation of ethane from ethylene and hydrogen is of the second order. The essential condition in the first-order equation is that the chain shall be terminated by a reaction between the two different carriers of the chain. In acetaldehyde the 1·5 order results from the fact that the chain is terminated by the reaction of two methyl groups. This difference is due to a difference in the heats of activation. The heats of activation can be determined from the measured reaction rates with considerable certainty, and it is seen that they fit the observed orders of the reaction, whereas assigning different values to the heats of activation might be expected to lead to a different order. Although there is as yet no experimental evidence for the presence of reaction chains in these decompositions and, further, there has been no direct experimental demonstration even of the presence of free radicals below 700° and above a few mm pressure, nevertheless it should be pointed out that the only way to avoid chain reactions as the explanation of the measured rates would be to increase the heats of activation involved to quite improbable values.

ON A SINGLE CHAIN MECHANISM FOR THE THERMAL DECOMPOSITION OF HYDROCARBONS†

V. A. POLTORAK and V. V. VOEVODSKY

[After the publication of Rice and Herzfeld's paper (Paper 11) it was generally assumed that organic decompositions occur largely by free-radical chain mechanisms. Some doubt was cast upon this, however, by the results of experiments made by L. A. K. Staveley and C. N. Hinshelwood (*Proc. Roy. Soc.* A, **154**, 335 (1936); *J. Chem. Soc.* 1568 (1937); cf. also L. A. K. Staveley, *Proc. Roy. Soc.* A, **162**, 557 (1937)) and by results of several subsequent workers. These experiments showed that when nitric oxide and other substances are added to decomposing organic compounds there is inhibition, but that a certain fraction of the reaction (often 10 per cent or more) could not be eliminated. This fraction was thought by Hinshelwood and others to be a purely molecular mechanism, but there were some difficulties with this interpretation.

In the following paper this problem is examined, and it is concluded that in the propane decomposition the residual reaction is not molecular. The same conclusion has subsequently been reached for other decompositions, such as that of ethane (F. O. Rice and R. E. Varnerin, *J. Amer. Chem. Soc.* **76**, 324 (1954)), butane (A. Kuppermann and L. G. Larson, *J. Chem. Phys.* **33**, 1264 (1960)) and dimethyl ether (D. J. McKenney, B. W. Wojciechowski and K. J. Laidler, *Can. J. Chem.* **41**, 1993 (1963)).

Vladislav Vladislavowich Voevodsky was born in 1917, and at the time this paper was written was associated with the Institute of Chemical Physics of the Academy of Sciences of the U.S.S.R. and with Moscow State University. From 1961 until his untimely death in 1967 he was Vice-Director of the Institute of Chemical Kinetics and Combustion in Novosibirsk, and simultaneously Professor of Chemistry in the Novosibirsk State University. He was a full Member (Academician) of the

---

† Translation of a paper published in *Doklady Academii Nauk S.S.S.R.*, **91**, 589 (1953).

Academy of Sciences of the U.S.S.R., and made many important
contributions in the field of chemical kinetics. Mrs. Valentina Aleksan-
drovna Poltorak was born in 1925 and at the time this paper was written
was an aspirant (graduate student) in the chemical kinetics laboratories
of Moscow State University. She is now in the Department of Biology of
Moscow State University.]

THE mechanism for the thermal decomposition of hydrocarbons
and their derivatives has recently been suggested, in some publica-
tions by foreign authors, as consisting of two concurrent processes
—free-radical chain and molecular (a direct redistribution of the
bonds in the initial molecule). The main argument for this assumption is based on the results of numerous experiments on the
inhibition of the decompositions by different compounds (nitric
oxide, propylene and higher olefins and other substances); on
adding these inhibitors it is possible to suppress only a part of the
reaction. Further additions of inhibitor, however, do not further
decrease the rate; in some cases they actually increase it signifi-
cantly. This was interpreted by Hinshelwood and other authors[1]
as indicating that in addition to a free-radical chain decomposition,
which can be easily suppressed by the reactive inhibitors, a purely
molecular decomposition also occurs, which accounts for an
appreciable fraction of the decomposition.

Some objections have been raised against this point of view, the
main one being that in different reacting systems (paraffins, alde-
hydes, ethers, etc.) it is very unlikely that over a wide range of
temperature and pressure the decompositions would always occur
with comparable rates by the two mechanisms, and result in
practically the same ratios of the final products. In particular,
Gol'danskii[2] has shown that the similar results can be explained
if the NO molecule acts in two different ways. If it is assumed that
NO catalyses both the initiation and termination of the chains, the
rate of the reaction may be considerably decreased without a

[1] F. J. Stubbs and C. N. Hinshelwood, *Proc. Roy. Soc.* A **200**, 458 (1950);
A **201**, 18 (1950); K. V. Ingold, F. J. Stubbs and C. N. Hinshelwood, *ibid.*,
A **203**, 486 (1950); A **208**, 285 (1951); A **214**, 20 (1952).
[2] V. I. Gol'dansky, *Usp. Kim.* **15**, 63 (1946).

change in the chain nature of the decomposition. Other explanations based entirely on a chain mechanism are also possible. Unambiguous experimental proof of the presence or absence of free radicals in the inhibited reaction thus became necessary.

For this purpose we decided to use a method developed in our laboratories for the detection of free radicals using molecular deuterium.[3] This method is based on the fact that alkyl radicals

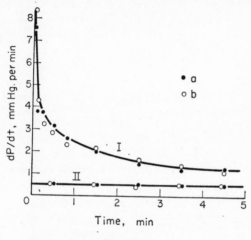

FIG. 12.1

$T = 570°C$

a   50 mm C_3H_8
b   40 mm H_2 + 50 mm C_3H_8
I   Rate of decomposition of C_3H_8
II   Rate of decomposition of C_3H_6 in the presence of 20% NO.

react rapidly with deuterium to form partially deuterated products. From the fraction of deuterium in the products it is possible to determine whether these products were formed from radicals or directly by a molecular decomposition.

The decomposition of propane was studied at the temperatures 570 and 600°C and with a pressure of propane of 50 mm. In the

[3] V. V. Voevodsky, G. K. Lavrovskaya and R. E. Mardaleyshvili, *Doklady Acad. Nauk* **81**, 215 (1951).

presence of NO it was found, in agreement with previous experiments, that the rate of reaction was practically constant up to a considerable extent of decomposition, whereas with pure propane the reaction is noticeably inhibited as the reaction proceeds (Fig. 12.1). Special experiments showed that the addition of 40 mm of hydrogen to the 50 mm of propane and to a mixture of 50 mm propane and $12 \cdot 5$ mm NO did not change the kinetics of the process. Experiments were then carried out with the following mixtures: 50 mm $C_3H_8 + 40$ mm D_2; 50 mm $C_3H_8 + 40$ mm $D_2 + 12 \cdot 5$ mm NO; 50 mm $C_3H_8 + 25$ mm D_2; 50 mm$C_3H_8 + 25$ mm $D_2 + 12 \cdot 5$ mm NO.

All experiments were carried out to 10% decomposition. The products ethylene and ethane, the latter not more than 5% of the products formed, were separated from the reaction mixture and converted to water by combustion over copper oxide at 600°C.

TABLE 1

PERCENTAGES OF D_2O IN THE WATER FORMED BY BURNING THE $C_2H_4 + C_2H_6$ FRACTION

Temperature (°C)	Composition of reaction mixture (mm Hg)	Per cent D_2O	Average per cent of D
570	$50C_3H_8 + 40D_2$	$3 \cdot 5$; $7 \cdot 2$; $5 \cdot 2$; $3 \cdot 5$	$4 \cdot 85 \pm 1 \cdot 3$
	$50C_3H_8 + 40D_2 + + 12 \cdot 5NO$	$7 \cdot 9$; $3 \cdot 3$; $3 \cdot 3$; $9 \cdot 4$; $3 \cdot 0$; $4 \cdot 0$	$5 \cdot 15 \pm 2 \cdot 3$
	$50C_3H_8 + 25D_2$	$2 \cdot 8$	$2 \cdot 8$
	$50C_3H_8 + 25D_2 + + 12 \cdot 5NO$	$2 \cdot 6$	$2 \cdot 6$
600	$50C_3H_8 + 40D_2$	$5 \cdot 6$; $5 \cdot 0$	$5 \cdot 3 \pm 0 \cdot 3$
	$50C_3H_8 + 40D_2 + + 12 \cdot 5NO$	$4 \cdot 7$; $3 \cdot 6$; $6 \cdot 2$; $9 \cdot 7$	$6 \cdot 05 \pm 1 \cdot 9$
	$50C_3H_8 + 25D_2$	$3 \cdot 1$	$3 \cdot 1$
	$50C_3H_8 + 25D_2 + + 12 \cdot 5NO$	$3 \cdot 0$	$3 \cdot 0$

The water was then analyzed for deuterium content. These results, shown in Table 1, show that the deuterium content in the products was the same in the reaction inhibited by NO as in the uninhibited reaction. The fact that the exchange in the ethylene $+$ ethane fraction does not depend on the time of the reaction (the ratio of the times necessary for 10% decomposition in the experiments with NO and without NO was 7:1), and that it is considerably lower than an average equilibrium percentage deuterium in the whole system, excludes the possibility of a molecular exchange reaction of the saturated hydrocarbon with the deuterium.

Our results show, therefore, that the thermal decomposition of propane is a free-radical process both in the pure gas and in the presence of NO. They disprove the hypothesis that there are two simultaneous mechanisms of thermal decomposition.

[There still remains the question of exactly what are the mechanisms of reactions that are maximally inhibited by nitric oxide. As referred to in this paper, Gol'danskii suggested that nitric oxide is concerned both in initiating and terminating chains. Detailed mechanisms based on the proposal that nitric oxide abstracts a hydrogen atom from the substrate molecule have been put forward by Wojciechowski and Laidler (*Can. J. Chem.* **38**, 1027 (1960); *Trans. Faraday Soc.* **59**, 369 (1963)). Voevodsky (*Akademia Nauk S.S.S.R.* **4**, 603 (1964)), on the other hand, has advanced a hypothesis based on heterogeneous initiation and termination of chains, the effect of the nitric oxide being to reduce the rate of the heterogeneous initiation. Alternatively R. G. W. Norrish and G. L. Pratt (*Nature*, **197**, 143 (1963)) have suggested that the behaviour with nitric oxide is to be explained in terms of the decomposition of oximes formed as intermediates. Further work is needed in order to resolve this problem.]